AF360743

Coatings Imparting Multifunctional Properties to Materials

Coatings Imparting Multifunctional Properties to Materials

Editor

Natalia Prorokova

MDPI • Basel • Beijing • Wuhan • Barcelona • Belgrade • Manchester • Tokyo • Cluj • Tianjin

Editor
Natalia Prorokova
G.A. Krestov Institute of
Solution Chemistry of the
Russian Academy of Sciences
Russia

Editorial Office
MDPI
St. Alban-Anlage 66
4052 Basel, Switzerland

This is a reprint of articles from the Special Issue published online in the open access journal *Coatings* (ISSN 2079-6412) (available at: https://www.mdpi.com/journal/coatings/special_issues/Coat_Imp_Multifunct).

For citation purposes, cite each article independently as indicated on the article page online and as indicated below:

LastName, A.A.; LastName, B.B.; LastName, C.C. Article Title. *Journal Name* **Year**, *Volume Number*, Page Range.

ISBN 978-3-0365-2448-1 (Hbk)
ISBN 978-3-0365-2449-8 (PDF)

Contents

About the Editor

Natalia Prorokova is the chief researcher in the laboratory "Chemistry and technology of modified fibrous materials" of the G.A. Krestov Institute of Solutions Chemistry of the Russian Academy of Sciences and Professor of the Department of Natural Sciences and Technosphere Safety at Ivanovo State Polytechnic University. Professor N. Prorokova's scientific interests include the modification of synthetic fibrous materials, functional nanoscale coatings, the bulk nanomodification of synthetic fibrous materials in the molding process, and the use of fluoropolymer materials in the processes of fiber modification. Professor N. Prorokova's scientific works have been presented at many international exhibitions held in Germany, China, Switzerland, and Russia. Based on their results, a number of awards were received, notably the Gold Medal of the Geneva International Innovation Salon "Innovations - Geneva", several silver and bronze medals at the "Archimedes" exhibitions (Moscow), and a large number of diplomas. Professor N. Prorokova is the author of more than 400 scientific works, including chapters in 8 monographs, 4 teaching and methodological manuals, and 12 patents.

Editorial

Special Issue "Coatings Imparting Multifunctional Properties to Materials"

Natalia Prorokova [1,2]

[1] G.A. Krestov Institute of Solution Chemistry of the Russian Academy of Sciences, Akademicheskaya St. 1, 153045 Ivanovo, Russia; npp@isc-ras.ru or npp238@gmail.com
[2] Department of Natural Sciences and Technosphere Safety, Ivanovo State Polytechnic University, Sheremetevsky Ave. 21, 153000 Ivanovo, Russia

Citation: Prorokova, N. Special Issue "Coatings Imparting Multifunctional Properties to Materials". *Coatings* **2021**, *11*, 1362. https://doi.org/10.3390/coatings11111362

Received: 1 November 2021
Accepted: 4 November 2021
Published: 5 November 2021

Publisher's Note: MDPI stays neutral with regard to jurisdictional claims in published maps and institutional affiliations.

Coating the surface of various materials and products has been used for a long time for protection against corrosion and erosion, in order to increase the service life and productivity of equipment. Another equally important area of application of coatings is to impart new properties to materials. It is known that the properties of the surface and near-surface layer largely determine the performance characteristics of materials and their processing ability. Therefore, the application of specially selected coatings is an effective tool for surface modification of materials to give them new and properties (for example, biocidal, hydrophobic, increased electrical conductivity or, on the contrary, dielectric properties, etc.). In recent decades, researchers have paid special attention to the creation of multifunctional coatings.

This book contains the results of the latest research on the formation of coatings on the surface of various materials. The composition of the coatings formed by researchers is very diverse. Coatings are selected depending on the set of properties that need to be given to the material being modified. However, a significant part of researchers choose composite materials based on various polymers as a material for the formation of strong and durable coatings with high adhesion to the substrate.

The articles presented in the book can be divided into two groups. The first one is devoted to the solution of the classical, but still very urgent problem of imparting superhydrophobicity to materials and other properties related to this characteristic. In particular, the article "Superhydrophobic Al_2O_3–polymer composite coating for self-cleaning applications" by R.S. Sutar with co-authors is dedicated to imparting superhydrophobic and self-cleaning properties to glass substrates [1]. The approach substantiated by the authors differs from numerous works on this topic by the fact that the researchers achieve extremely high indicators of glass hydrophobicity by applying a coating manage. The article "Robust superhydrophobic and repellent coatings based on micro/nano SiO_2 and fluorinated epoxy" by X. Huang and R. Yu presents the development of a universal method for imparting high hydrophobic properties to materials of all kinds [2]. A serious achievement of the authors of the study is also a high resistance of the coating to mechanical and chemical influences, which is usually difficult to ensure. An important place in the book is occupied by the review article "Critical aspects in fabricating multifunctional super-nonwettable coatings exhibiting icephobic and anti-biofouling properties" by K.D. Esmeryan, which analyzes the most significant, according to the author, publications over the past five years, which are devoted to the formation of coatings with superhydrophobic and ice-phobic properties that prevent biofouling [3].

The second group of articles presented in the book is devoted to the formation of coatings that provide various fibrous materials with a large group of special properties. These articles differ both in the objects of modification (cellulose textile material, adhesive interlining materials for clothes, polypropylene yarn), and in the methods of forming the coatings used in them.

L. Petrova and co-authors, in the article "Development of multifunctional coating of textile materials using silver microencapsulated compositions", apply a well-known layer-by-layer method [4]. N. Kornilova and co-authors, in the article "Multifunctional polymer coatings of fusible interlinings for sewing products", use a popular method of dipping to obtain coatings [5]. The article "Properties of polypropylene yarns with a polytetrafluoroethylene coating containing stabilized magnetite particles" by N. Prorokova and S. Vavilova develops a new method of forming coatings based on the application of a coating-forming composition on the surface of a semi-cured thermoplastic filament at the stage of its melt spinning followed by orientational stretching [6]. As a result of coating deposition, fibrous materials acquire a set of new properties that are of great practical importance. Cellulose fibrous materials become antibacterial, antimycotic, and gain wound healing properties. Interlining materials for clothing acquire shape stability, wear resistance, and protective and health-improving properties at the same time. Polytetrafluoroethylene-coated polypropylene yarn is characterized by good antimicrobial properties, reduced electrical resistance, increased strength, and chemical resistance.

This book presents only a small part of the current research devoted to the formation of coatings to impart multifunctional properties to materials. This topic is one of the most relevant areas of modern materials science and, undoubtedly, will be actively developed.

Funding: This research received no external funding.

Institutional Review Board Statement: Not applicable.

Informed Consent Statement: Not applicable.

Data Availability Statement: Not applicable.

Conflicts of Interest: The author declares no conflict of interest.

References

1. Sutar, R.S.; Nagappan, S.; Bhosale, A.K.; Sadasivuni, K.K.; Park, K.-H.; Ha, C.-S.; Latthe, S.S. Superhydrophobic Al_2O_3–Polymer Composite Coating for Self-Cleaning Applications. *Coatings* **2021**, *11*, 1162. [CrossRef]
2. Huang, X.; Yu, R. Robust Superhydrophobic and Repellent Coatings Based on Micro/Nano SiO_2 and Fluorinated Epoxy. *Coatings* **2021**, *11*, 663. [CrossRef]
3. Esmeryan, K.D. Critical Aspects in Fabricating Multifunctional Super-Nonwettable Coatings Exhibiting Icephobic and Anti-Biofouling Properties. *Coatings* **2021**, *11*, 339. [CrossRef]
4. Petrova, L.; Kozlova, O.; Vladimirtseva, E.; Smirnova, S.; Lipina, A.; Odintsova, O. Development of Multifunctional Coating of Textile Materials Using Silver Microencapsulated Compositions. *Coatings* **2021**, *11*, 159. [CrossRef]
5. Kornilova, N.; Bikbulatova, A.; Koksharov, S.; Aleeva, S.; Radchenko, O.; Nikiforova, E. Multifunctional Polymer Coatings of Fusible Interlinings for Sewing Products. *Coatings* **2021**, *11*, 616. [CrossRef]
6. Prorokova, N.; Vavilova, S. Properties of Polypropylene Yarns with a Polytetrafluoroethylene Coating Containing Stabilized Magnetite Particles. *Coatings* **2021**, *11*, 830. [CrossRef]

Article

Superhydrophobic Al₂O₃–Polymer Composite Coating for Self-Cleaning Applications

Rajaram S. Sutar [1], Saravanan Nagappan [2,3], Appasaheb K. Bhosale [1,*], Kishor Kumar Sadasivuni [4], Kang-Hyun Park [2], Chang-Sik Ha [3] and Sanjay S. Latthe [1,*]

1   Self-cleaning Research Laboratory, Department of Physics, Raje Ramrao College, Jath, Affiliated to Shivaji University, Kolhapur 416404, Maharashtra, India; raju.sutar2@gmail.com

2   Department of Chemistry, Chemistry Institute for Functional Materials, Pusan National University, Busan 46241, Korea; saravananagappan@gmail.com (S.N.); chemistry@pusan.ac.kr (K.-H.P.)

3   Department of Polymer Science and Engineering, Pusan National University, Busan 46241, Korea; csha@pusan.ac.kr

4   Center for Advanced Materials, Qatar University, Doha P.O. Box 2713, Qatar; chishorecumar@gmail.com

*   Correspondence: akbhosale1@gmail.com (A.K.B.); latthes@gmail.com (S.S.L.)

Abstract: Superhydrophobic coatings have a huge impact in various applications due to their extreme water-repellent properties. The main novelty of the current research work lies in the development of cheap, stable, superhydrophobic and self-cleaning coatings with extreme water-repellency. In this work, a composite of hydrothermally synthesized alumina (Al₂O₃), polymethylhydrosiloxane (PMHS) and polystyrene (PS) was deposited on a glass surface by a dip-coating technique. The Al₂O₃ nanoparticles form a rough structure, and low-surface-energy PHMS enhances the water-repellent properties. The composite coating revealed a water contact angle (WCA) of $171 \pm 2°$ and a sliding angle (SA) of $3°$. In the chemical analysis, Al2p, Si2p, O1s, and C1s elements were detected in the XPS survey. The prepared coating showed a self-cleaning property through the rolling action of water drops. Such a type of coating could have various industrial applications in the future.

Keywords: alumina (Al₂O₃) coating; superhydrophobic; self-cleaning; composite coating

Citation: Sutar, R.S.; Nagappan, S.; Bhosale, A.K.; Sadasivuni, K.K.; Park, K.-H.; Ha, C.-S.; Latthe, S.S. Superhydrophobic Al₂O₃–Polymer Composite Coating for Self-Cleaning Applications. *Coatings* **2021**, *11*, 1162. https://doi.org/10.3390/coatings11101162

Academic Editor: N. P. Prorokova

Received: 27 August 2021
Accepted: 22 September 2021
Published: 27 September 2021

Publisher's Note: MDPI stays neutral with regard to jurisdictional claims in published maps and institutional affiliations.

1. Introduction

Superhydrophobic surfaces have earned much attention from researchers in the last two decades due to their excellent water repellent behavior and the high mobility of water, which can be used to avoid accumulating dirt, fouling, fogging, and icing [1–4]. Natural leaf surfaces, such as that of a lotus leaf, possess micro-scale papillae and nano-scale epicuticular wax crystals on their surface, forming a hierarchical surface morphology, which is responsible for their self-cleaning superhydrophobic properties by quickly removing the dirt particles from the surface by rolling water drops [5]. To date, many efforts have been devoted to the development of superhydrophobic coatings by forming a rough structure and/or reducing the surface energy by using low surface energy materials [6–10]. The use of low surface energy-based materials on the rough hierarchical structure may create a thin hydrophobic layer that could resist the adherence of water droplets [8,11,12]. The presence of a hierarchical surface with a thin layer of hydrophobic materials may also be another reason for the superhydrophobic property [13]. Similarly, different kinds of bio-mimicking surfaces were developed using natural or synthetic materials to achieve a micro-nano hierarchical surface structure with an extreme water repellent coating for self-cleaning, as well as oil–water sorption and separation applications [14,15]. Several studies were studied a mechanism of superhydrophobic as well as photocatalytic superhydrophilic surfaces in self-cleaning applications [16,17]. Photocatalytic superhydrophilic surfaces have also attracted considerable attention in terms of self-cleaning coatings due to the complete wettability of their substrates, which can easily remove an organic pollutant from the surfaces by the action of the flow of a water film [16,17]. Esmeryan et al. developed a novel soot-inspired superhydrophobic surface containing a quartz crystal microbalance (QCM)-based biosensor for the detection of human semen, and human spermatozoa quality assessment [18]. Similarly, the authors also studied the effect of soot-inspired superhydrophobic surfaces for human urine detection, as well as improving the success rate of the cryopreservation of human spermatozoa [19,20]. Here, we only focused on the fabrication of superhydrophobic coatings for self-cleaning applications.

Aluminium oxide (Al_2O_3) is one of the cheapest materials, with excellent usability in various applications. The use of Al_2O_3 in coatings also attracted significant attention in recent years due to its antibacterial property; excellent mechanical, electrical insulation, high-temperature properties; and first-rate impact, abrasion, and chemical resistance [21]. Several studies have focused on developing superhydrophobic surfaces using Al_2O_3 particles or on the Al_2O_3 surface [22–25]. Sutha et al. fabricated an optically transparent, anti-reflective, and self-cleaning superhydrophobic Al_2O_3 coating on a glass substrate [25]. In this process, the authors first prepared Al_2O_3 sol by mixing aluminium nitrate non-ahydrate with 2-methoxyethanol solution by magnetic stirring in a monoethanolamine stabilizer at room temperature. Multiple layers of Al_2O_3 nanoparticles were applied onto a glass substrate by a spin-coating method. After annealing, the film was immersed in hot water to obtain a porous structure. Finally, low surface energy 1H,1H,2H,2H–perfluorooctyltrichlorosilane was coated onto a porous Al_2O_3 film by spin coating. On the other hand, Karapanagiotis et al. dispersed different-sized hydrophilic alumina nanoparticles (25, 35, and 150 nm) in different concentrations in solutions of a hydrophobic poly(alkyl siloxane), and the prepared suspensions were sprayed onto a glass surface [26]. They stated that the wettability of the composite film is independent of the size, but is affected by the concentration of the particles. However, Richard et al. dispersed stearic acid modified Al_2O_3 particles in ethanol, and sprayed them onto a glass slide in order to attain a superhydrophobic surface [27]. Tie et al. prepared a superhydrophobic and underwater superoleophobic surface by an aqueous mixture of hydrophilic nanoparticles (TiO_2, SiO_2, and Al_2O_3) and fluorocarbon surfactants through dip, brush, or spray coating on various substrates, such as fabric, sponge, cotton, nickel foam, stainless steel mesh, copper sheet, glass, and ceramics [28]. Byun et al. prepared a superhydrophobic surface by spraying phosphonic acid-functionalized Al_2O_3 nanoparticles onto glass, paper, cotton fabric, and flexible plastic substrates [29]. Several studies are available on the fabrication of superhydrophobic surfaces with extreme wettability by different techniques [13,14].

Although several techniques were used to fabricate superhydrophobic surfaces with self-cleaning behavior using various materials, only a few works were reported using Al_2O_3 nanoparticle-based nanocomposites for superhydrophobic and self-cleaning coatings [22,30,31]. Al_2O_3 based composites are highly useful in coating applications due to the abundant availability of the Al source and its antibacterial characteristics [32]. In the present work, we exclusively focused on a facile dip-coating method to coat an Al_2O_3–PMHS–PS hybrid system onto glass substrates. First, hydrothermally synthesized hydrophilic Al_2O_3 nanoparticles were modified with low-surface-energy PMHS. The flake-shaped Al_2O_3 nanoparticles were agglomerated during the deposition, forming a rough hierarchical structure and attaining a superhydrophobic surface. The chemical analysis of the prepared coating was also performed. Water jet impact, adhesive tape peeling and sandpaper abrasion tests were performed in order to evaluate the mechanical durability of the coating. Additionally, self-cleaning tests were conducted on the prepared superhydrophobic coatings. The coated substrates exhibited extreme water-repellent behavior as well as excellent mechanical stability.

2. Experiment

2.1. Materials

Aluminum nitrate nonahydrate [$Al(NO_3)_3 \cdot 9H_2O$], polystyrene (PS; 192,000 g/mol) and polymethylhydrosiloxane (PMHS, average Mn 1700-3200) were procured from Sigma-Aldrich

(St. Louis, MO, USA). Dextrose [O(CHOH)$_4$CHCH$_2$OH] and urea [NH$_2$CONH$_2$] were secured from Thomas Baker (Mumbai, India). Ethanol and chloroform were bought from Spectrochem PVT. LTD (Mumbai, India). The micro-Glass substrates ($75 \times 25 \times 1.35$ mm^3) were obtained from Blue star, Polar Industrial Corporation, Mumbai, India.

2.2. Synthesis of the Al$_2$O$_3$ Nanoparticles

Aluminum nitrate nonahydrate has been used to synthesize Al$_2$O$_3$ nanoparticles via a hydrothermal method [33,34]. In the synthesis process, 3.6 g dextrose, 3.75 g aluminum nitrate nonahydrate, and 3 g urea were dissolved in 50 mL distilled water under vigorous stirring for 30 min. The prepared homogeneous transparent solution was transferred to a 100 mL Teflon-lined stainless-steel autoclave and kept in an oven at 170 °C for 6 h for the hydrothermal process. A black powder of hydrated alumina was collected by filtration and washed with distilled water and ethanol several times, and was dried at 80 °C for 8 h. The dried black powder of hydrated alumina was placed in a silica crucible and kept in a muffle furnace at 1000 °C for 3 h, with a heating rate of 4 °C min^{-1} in air atmosphere. Finally, the collected Al$_2$O$_3$ particles were stored in the bottle for further use.

2.3. Preparation of the Superhydrophobic Coating

The micro-glass slides were washed using tap water and laboratory detergent (Moly-clean 02 Neutral, Molychem, Mumbai, India) and, afterwards, cleaned ultrasonically using distilled water and ethanol for 10 min. A coating solution was prepared by the following process: 0.15 mL PMHS was mixed with 20 mL chloroform in a beaker, and kept on magnetic stirrer at 100 rpm. After 20 min of stirring, 400 mg Al$_2$O$_3$ was added and stirred continuously for another 1 h. Meanwhile, in a second beaker, 10 mg/mL PS solution in 20 mL of chloroform was prepared. This PS solution was poured into the beaker containing PMHS-Al$_2$O$_3$ and stirred further for 30 min.

The cleaned glass slide was dipped in a suspension of Al$_2$O$_3$–PMHS–PS for 10 s, with a controlled dip and withdrawal rate of 50 mm/s using a dip coating machine (Delta Scientific Equipment Pvt. Ltd., Kolkata, India); a coated sample was dried at room temperature (~25 °C). This is considered as one deposition layer of the coating solution. The coatings were repeated by applying 2, 4, and 6 deposition layers, and finally dried at 100 °C in an oven for 1 h. The samples with two, four, and six coating layers were labelled as the AD-1, AD-2 and AD-3 coating, respectively. In this work, we only focused on the study of the effect of dip-coating cycles on the superhydrophobic coating. Of course, changing the chemical compositions would alter the surface properties either partially or completely based on the formation of a hierarchical surface morphology, with more hydrophobic or hydrophilic characteristics based on the combinations of material.

2.4. Characterizations

A Scanning Electron Microscope (SEM, JEOL, JSM-7610F, Tokyo, Japan) was used to investigate the surface micro/nanostructure of the prepared coatings. The surface roughness was calculated using a Stylus profiler (Mitutoyo, SJ 210, Sakado, Japan). The water contact angle (WCA) and sliding angle (SA) were measured on at least three places on the samples using a contact angle meter (HO-IAD-CAM-01, Holmarc Opto-Mechatronics Pvt. Ltd., Kochi, India). The average value of the WCA and SA of samples were noted. The chemical composition of the coating was analyzed by X-ray photoelectron spectroscopy (XPS, PHI Quantera-II, Tokyo, Japan). The mechanical stability of the coatings was examined using a water jet created by a syringe on the coating. The mechanical sustainability of the coatings was evaluated further by an adhesive tape test and a sandpaper abrasion test using commercial adhesive tape and sandpaper. The self-cleaning performance of the coatings was observed by scattering fine particles of chalk as a dust contaminant onto the coating.

3. Results and Discussion

3.1. Surface Morphology, Roughness and Wettability of the Prepared Coatings

The surface micro/nanostructure has been given much attention in the definition of the wetting property of the coating surface. Mostly, micro- and nano-scale hierarchical surface structures with low surface energy are responsible for superhydrophobicity. The surface morphologies of the prepared coatings was analyzed by SEM, and are given in Figure 1a–f. The flake-shaped alumina nanoparticles formed during the hydrothermal synthesis are visible in the SEM images [35]. The addition of PMHS and PS molecules to the flake-shaped nanoparticles can form an aggregated micro–nano-sized random particles during the deposition process. Moreover, the agglomerated particles provided micro- and nano-sized hierarchical rough structures on the glass surface. The active Si-H bond and methyl groups present in the PMHS were utilized in the surface modification of hydrophilic alumina nanoparticles. The formation of a thin layer of hydrophobic PMHS on Al_2O_3 would facilitate the enhancement of the hydrophobic property on the modified Al_2O_3–PMHS surface. Radwan et al. reported that the mixture of PS and Al_2O_3 can deliver the superhydrophobic property while forming three dimensional nanofibers by an electrospinning technique [22]. PS nanofibers can display an excellent hydrophobic property, which becomes superhydrophobic through the introduction of Al_2O_3 to the PS due to the formation of a multiscale hierarchical rough structure [22]. Likewise, the modification of PS with various hydrophobic agents can also develop a superhydrophobic surface property [36–38]. As such, the addition of a PS solution to Al_2O_3–PMHS would also help to form the multiscale hierarchical roughness by aggregation, as well as the formation of closely packed particles on the coated substrate.

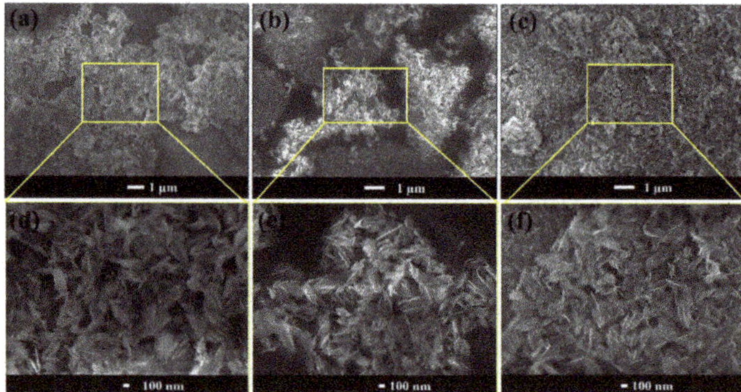

Figure 1. (**a–c**) Low-magnified and (**d–f**) high-magnified FE-SEM images of the AD-1, AD-2 and AD-3 samples, respectively.

At two layers (AD-1) of deposition, the particles were agglomerated and uniformly distributed on the glass surface, as shown in Figure 1a,d. The AD-1 coating showed a surface roughness of 0.019 μm with WCA 120 ± 2°, and a water drop becomes stuck on the surface. The size of the agglomerated particles increased with increasing numbers of layers, up to four (AD-2), resulting in a highly rough structure with a roughness value of 0.038 μm, which is similar to a Cassie-Baxter surface. The developed micro/nano-sized rough structure of the coating is clearly seen in Figure 1b,e. In such a hierarchical surface structure, air pockets are trapped; consequently, WCA increased to 171 ± 2°, and water drops roll off at an inclination angle of 3 ± 0.5°. Based on the formation of a multiscale micro-nano hierarchical structure as well as the formation of thin layer of low surface energy hydrophobic PMHS, this provides an extreme superhydrophobic property on a dip-coated substrate [39]. Further increasing the number of deposition layers to six, we noticed that more agglomerated particles are formed on the glass surface, and the surface roughness decreased partially to 0.034 μm (Figure 1c,f). At the same time, the six-layer coated sample also exhibited WCA 170 ± 2°, with no changes in sliding angles. The AD-3 coated substrate can also deliver good mechanical durability when it forms densely packed particles on the coated surface, whereas the loosely packed particles on the surface mean that it can easily come out from the substrate under adhesive tape peeling and sandpaper abrasion tests [40]. As such, we further studied the effect of an AD-2-coated superhydrophobic substrate for the rest of the studies, because at four layers of coating, the fabricated substrate can exhibit the maximum contact angle, as well as a surface roughness which was reduced by the further increase of coating layers. As such, considering the practical point of view as well industrial applications, four layers of coating would be the optimum in order to minimize the time consumption of the coated solution.

The typical photographs of Al_2O_3–PMHS–PS composite-coated substrates are shown in Figure 2. A coated substrate has a translucent or opaque color due to the deposition of white Al_2O_3 and PS on the glass substrate. The PMHS solution was transparent, and its addition was helpful for a stronger adhesion of the composite coating the substrate. The extreme superhydrophobic property of the AD-2 sample was confirmed by placing water droplets on the coated substrate. Figure 2a shows spherical-shaped color-dyed water drops, which explain the exceptional water-repellent behavior of the coated substrate. The inset of Figure 2a reveals an optical image of the water drop (approximately 10 µL volume) on a superhydrophobic AD-2 coating, which is obtained from a contact-angle meter. A water drop rolled off when the substrate was inclined by nearly 3° with the help of the stage of the contact-angle meter. Figure 2(c1–c3) shows the rolling action of a water drop on a superhydrophobic coating. The water drops quickly rolled down from an inclined surface, as shown in the inset of Figure 2(c3). The stability of the coating against running water was checked using a water jet hitting test. The water jet was formed using a 10 mL syringe, hitting a specific place on the coating for more than one minute (Figure 2b). The continuous reflecting water jet from the superhydrophobic surface confirmed that the prepared coating is highly stable. More results on the mechanical durability of the coating are provided in Section 3.3. Unfortunately, we are unable to provide an alternative image to showcase the sliding angle. Because of the extreme water repellent behavior of the coated superhydrophobic surface, as well as the very low sliding angle, a droplet can easily run away after contacting the substrate. Moreover, the inset image of c3 also clearly suggests the non-adherence of the water droplet after the transportation of the water droplet from the substrate. As such, we hope this image is enough to illustrate the low sliding angle and extreme water-repellency of the superhydrophobic substrate.

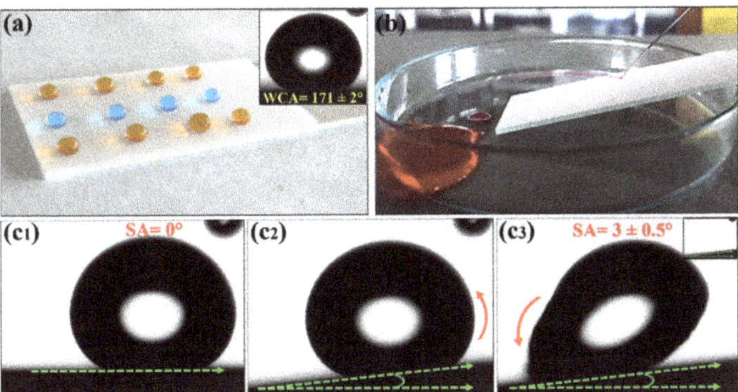

Figure 2. (a) Photograph of the color-dyed water drops on the superhydrophobic AD–2 coating. (b) The screenshot of the water jet hitting the superhydrophobic AD–2 coating. (c1–c3) The screenshots of the water droplet rolling on the superhydrophobic AD–2 coating.

3.2. XPS Study

The surface elemental compositions of the prepared superhydrophobic Al_2O_3-PMHS-PS composite AD-2 coating were analyzed using XPS studies (Figure 3a). Four peaks were observed at 74.85 eV, 102.61 eV, 284.8 eV and 532.11 eV, and are related to Al2p, Si2p, C1s, and O1s, respectively. The presence of Al, Si, C and O elements confirms that the Al_2O_3-PMHS-PS composite exists on the glass substrate. In Al2p scan spectra (Figure 3b), the peaks correspond to Al-O (74.6 eV) and Al-O-Si (75.5 eV) bonds. In Figure 3c, the Si2p peak at 102.4 eV and 103.7 eV corresponds to the Si-O-Si and Si-O-Al bonds [41]. In the O1s scan (Figure 3d), the highest peak is associated to Al-O-Si (532.7 eV) and Al-O-Al (531.1 eV) bonds, corresponding to the PMHS chain and Al_2O_3 nanoparticles [42,43]. In the high-resolution C1s XPS spectrum (Figure 3e), the BEs of 284.2 to 286.1 eV are related C–C/C–H and C=O, respectively.

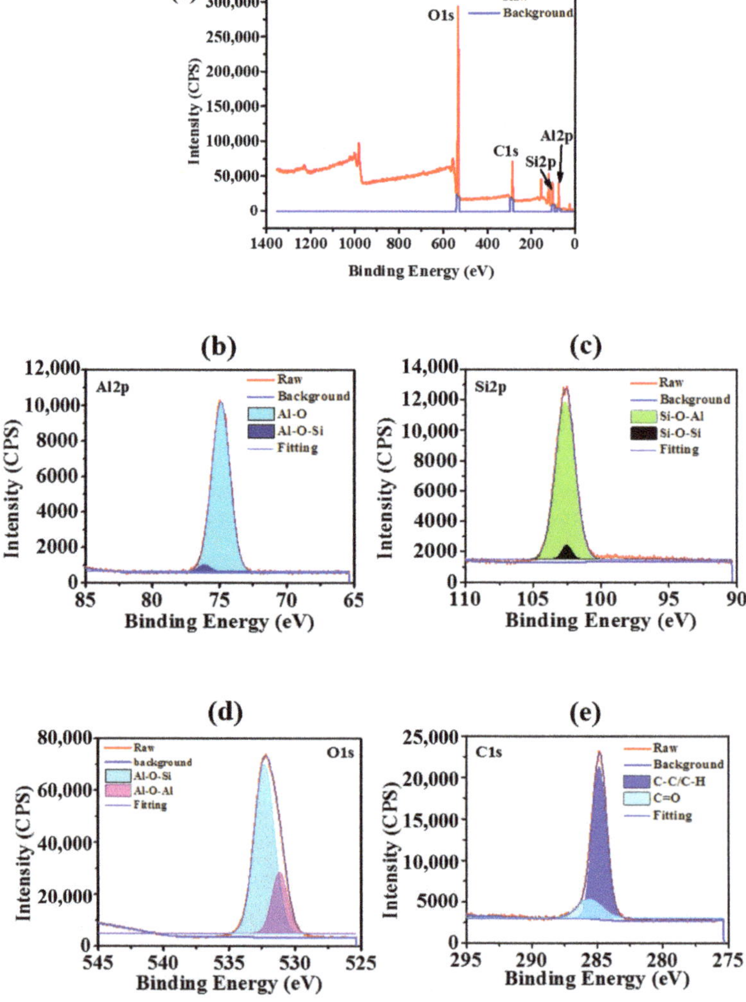

Figure 3. (**a**) XPS survey scan spectra of the superhydrophobic AD-2 coating. (**b**) Al2p peak, (**c**) Si2p peak, (**d**) C1s peak, and (**e**) O1 peak, respectively.

3.3. Mechanical Durability Tests

The mechanical durability of the prepared superhydrophobic coating is highly important for commercial applications. A fragile hierarchical structure of superhydrophobic coatings can be ruined when exposed to outdoor applications. The preparation of a robust superhydrophobic property is always a challenging issue because the superhydrophobic surface property can be damaged under severe mechanical stress, as well as under hot water or acidic and basic conditions. Several studies focused on improving the robustness of superhydrophobic surfaces by introducing highly strong adhesives.

The adhesive tape peeling and sandpaper abrasion tests are the most commonly used methods to evaluate the mechanical durability of superhydrophobic coatings [44]. A Cello tape no.405 (adhesiveness 3.93 N/10 mm) was placed on the AD-2 coating, and a metal disc of weight of 200 g was rolled on it to create good contact between the coating's surface and the tape. The tape was peeled off slowly from the coating's surface to check the adhesive tape peeling test performance, and this is considered one cycle of the tape-peeling test [44]. The WCA was also measured after test to check wetting property of the coating. The analysis of the WCA versus the number of tape peeling tests revealed that the superhydrophobicity remained stable for up to five cycles of tape peeling. After seven cycles, the WCA decreased to lower than ~140°. A variation of the WCA with an increasing number of tape-peeling cycles is shown in Figure 4a, and a photograph of the tape-peeling test is shown in the inset of Figure 4a.

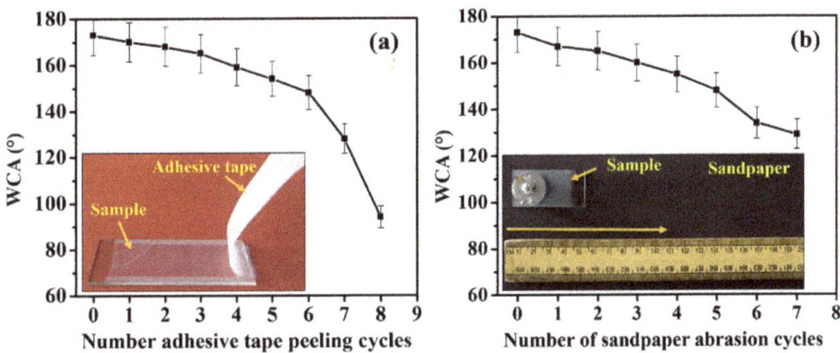

Figure 4. (**a**) An adhesive tape peeling test and (**b**) a sandpaper abrasion test on the AD-2 coating.

The AD-2 sample was placed on sandpaper of grit no. 400, then a 50 g weight was loaded onto it, and subsequently the sample was rubbed with a speed of ~5 mm/s for 10 cm (one cycle of the sandpaper abrasion test). A WCA was recorded after every sandpaper abrasion cycle (Figure 4b). The experimental setup of the sandpaper abrasion test is shown in the inset of Figure 4b. The WCA was reduced to ~162°, and the SA slightly increased (7°) after the completion of three cycles of the sandpaper abrasion test. This result indicated that the prepared coating is highly stable. After five abrasion cycles, the WCA decreased to ~150°. Subsequently, at seven cycles, water drops start to spread on the coating surface, as the composite material might have been removed by abrasion. A weak van der Waals force of attraction and hydrogen bonding can occur between the Al_2O_3 and PMHS, whereas the addition of PS to the Al_2O_3-PMHS suspension can have a hydrophobic–hydrophobic interaction between the hydrophobic PMHS and PS. While coating Al_2O_3-PMHS-PS composites onto a hydrophilic glass substrate, it would show a stronger interaction with the hydrophilic Al_2O_3. At the same time, an aggregated hydrophobic micro–nano hierarchical structure was observed on the surface, which produced a highly stable superhydrophobic property on the coated substrate under the thermal curing at 100 °C. These results suggest that the prepared Al_2O_3–PMHS–PS composite-coated substrate has excellent mechanical durability and robustness, which are important for practical applications. The coated substrate can maintain the superhydrophobic property for up to 5 to 6 cycles of adhesive tape peeling and sandpaper abrasion tests. We hope this mechanical durability is quite enough for various applications. Most of the superhydrophobic coatings reported can last less than five cycles of adhesive tape peeling and sandpaper abrasion tests [45–47]. As such, we hope that our coated substrate is better than the reported superhydrophobic coatings. Of course, enhanced robustness with the maintenance of superhydrophobicity and self-cleaning properties over 10 cycles of adhesive tape peeling and sandpaper abrasion tests are particularly recommended for high-end product development.

3.4. Self-Cleaning Test

Self-cleaning is one of the most desirable properties of a superhydrophobic coating. Such coatings can easily clean dust particles from their surface by the action of rolling water drops or without external force. Figure 5a shows a spreading of fine particles of colored chalk that are randomly scattered on the prepared superhydrophobic AD-2 coating, which was kept at an inclination angle nearly 10°. When a water shower produced by the syringe was sprinkled on this coating surface, owing to the highly water repellent property, the water drops rolled down the surface by collecting dust particles from the coating's surface (Figure 5b). Figure 5c illustrates that rolling water drops completely remove dust particles from the superhydrophobic surface, supporting the excellent self-cleaning property of the Al_2O_3–PMHS–PS composite coating. The superhydrophobic and self-cleaning properties are retained on the coated substrate by repeated wetting, and are also more durable in nature.

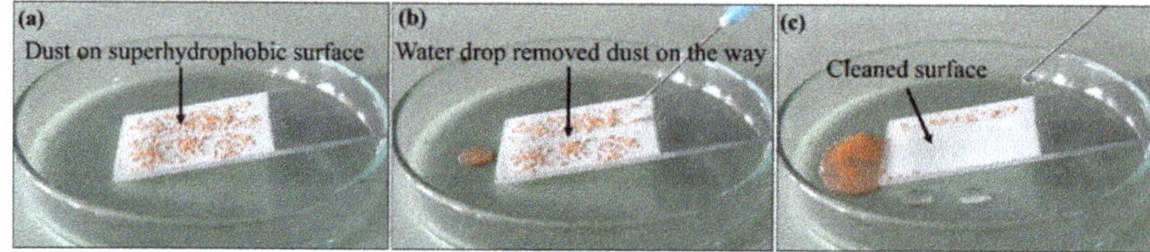

Figure 5. Self-cleaning performance of the prepared superhydrophobic AD-2 coating. (a) Dust particles on the superhydrophobic coating, (b) water drops carrying the dust particles, and (c) the cleaned surface.

4. Conclusions

We demonstrated a facile dip-coating method for the fabrication of a stable superhydrophobic coating. A stable superhydrophobic surface was achieved by the four-time dip coating of an Al_2O_3-PHMS-PS composite onto a glass substrate. The coated substrate presented a WCA ~171° and an SA ~3°. SEM micrographs of the coating showed flake-shaped alumina particles which were agglomerated and formed a rough microstructure. The XPS analysis revealed the co-existence of Al_2O_3, PHMS, and PS on the surface. The durability tests—such as the water jet impact, adhesive tape, and sandpaper abrasion test—displayed the high mechanical stability of the coating. In addition, the superhydrophobic Al_2O_3-PHMS-PS composite coating revealed excellent self-cleaning performance. The overall results suggest that the prepared Al_2O_3-PHMS-PS composite coating can be used to develop excellent superhydrophobic surfaces which might be potentially useful for various applications.

Author Contributions: Methodology and investigation—R.S.S.; review and modify the manuscript—S.N. and S.S.L.; supervision—A.K.B., K.K.S., K.-H.P., C.-S.H. and S.S.L. All authors have read and agreed to the published version of the manuscript.

Funding: This research was funded by Department of Science and Technology (DST), Goernment of India. [DST/INSPIRE/04/2015/000281] and National Natural Science Foundation of China (21950410531).

Institutional Review Board Statement: Not applicable.

Informed Consent Statement: Not applicable.

Data Availability Statement: Not applicable.

Acknowledgments: This work was financially supported by the DST–INSPIRE Faculty Scheme, Department of Science and Technology (DST), Goernment of India. [DST/INSPIRE/04/2015/000281]. SSL acknowledges financial assistance from Henan University, Kaifeng, China. We greatly appreciate the support of the National Natural Science Foundation of China (21950410531).

Conflicts of Interest: The authors declare no conflict of interest.

References

1. Latthe, S.S.; Sutar, R.S.; Kodag, V.S.; Bhosale, A.K.; Kumar, A.M.; Kumar Sadasivuni, K.; Xing, R.; Liu, S. Self-cleaning superhydrophobic coatings: Potential industrial applications. *Prog. Org. Coat.* **2019**, *128*, 52–58. [CrossRef]
2. Pan, R.; Zhang, H.; Zhong, M. Triple-Scale Superhydrophobic Surface with Excellent Anti-Icing and Icephobic Performance via Ultrafast Laser Hybrid Fabrication. *ACS Appl. Mater. Interfaces* **2021**, *13*, 1743–1753. [CrossRef] [PubMed]
3. Varshney, P.; Lomga, J.; Gupta, P.K.; Mohapatra, S.S.; Kumar, A. Durable and regenerable superhydrophobic coatings for aluminium surfaces with excellent self-cleaning and anti-fogging properties. *Tribol. Int.* **2018**, *119*, 38–44. [CrossRef]
4. Wang, D.; Sun, Q.; Hokkanen, M.J.; Zhang, C.; Lin, F.-Y.; Liu, Q.; Zhu, S.-P.; Zhou, T.; Chang, Q.; He, B. Design of robust su-perhydrophobic surfaces. *Nature* **2020**, *582*, 55–59. [CrossRef]
5. Dalawai, S.P.; Saad Aly, M.A.; Latthe, S.S.; Xing, R.; Sutar, R.S.; Nagappan, S.; Ha, C.S.; Kumar Sadasivuni, K.; Liu, S. Recent Advances in durability of superhydrophobic self-cleaning technology: A critical review. *Prog. Org. Coat.* **2020**, *138*, 105381. [CrossRef]
6. Geyer, F.; D'Acunzi, M.; Sharifi-Aghili, A.; Saal, A.; Gao, N.; Kaltbeitzel, A.; Sloot, T.-F.; Berger, R.; Butt, H.-J.; Vollmer, D. When and how self-cleaning of superhydrophobic surfaces works. *Sci. Adv.* **2020**, *6*, eaaw9727. [CrossRef] [PubMed]
7. Ma, W.; Ding, Y.; Zhang, M.; Gao, S.; Li, Y.; Huang, C.; Fu, G. Nature-inspired chemistry toward hierarchical superhydro-phobic, antibacterial and biocompatible nanofibrous membranes for effective UV-shielding, self-cleaning and oil-water separation. *J. Hazard. Mater.* **2020**, *384*, 121476. [CrossRef]
8. Teisala, H.; Butt, H.-J. Hierarchical Structures for Superhydrophobic and Superoleophobic Surfaces. *Langmuir* **2018**, *35*, 10689–10703. [CrossRef]
9. Wen, F.; Lei, C.; Chen, J.; Huang, Y.; Wang, B. Hierarchical superhydrophobic surfaces for oil–water separation via a gradi-ent of ammonia content controlling of dopamine oxidative self-polymerization. *J. Appl. Polym. Sci.* **2019**, *136*, 48044. [CrossRef]
10. Wen, R.; Xu, S.; Zhao, D.; Lee, Y.-C.; Ma, X.; Yang, R. Hierarchical Superhydrophobic Surfaces with Micropatterned Nan-owire Arrays for High-Efficiency Jumping Droplet Condensation. *ACS Appl. Mater. Interfaces* **2017**, *9*, 44911–44921. [CrossRef]
11. Kota, A.K.; Kwon, G.; Tuteja, A. The design and applications of superomniphobic surfaces. *NPG Asia Mater.* **2014**, *67*, e109. [CrossRef]
12. Si, Y.; Guo, Z. Superhydrophobic nanocoatings: From materials to fabrications and to applications. *Nanoscale* **2015**, *7*, 5922–5946. [CrossRef]
13. Shirtcliffe, N.J.; McHale, G.; Atherton, S.; Newton, M.I. An introduction to superhydrophobicity. *Adv. Colloid Interface Sci.* **2010**, *161*, 124–138. [CrossRef]
14. Nagappan, S.; Ha, C.-S. Emerging trends in superhydrophobic surface based magnetic materials: Fabrications and their po-tential applications. *J. Mater. Chem. A* **2015**, *3*, 3224–3251. [CrossRef]
15. Nagappan, S.; Park, J.J.; Park, S.S.; Lee, W.-K.; Ha, C.-S. Bio-inspired, multi-purpose and instant superhydrophobic–superoleophilic lotus leaf powder hybrid micro–nanocomposites for selective oil spill capture. *J. Mater. Chem. A* **2013**, *1*, 6761–6769. [CrossRef]
16. Adachi, T.; Latthe, S.S.; Gosavi, S.W.; Roy, N.; Suzuki, N.; Ikari, H.; Kato, K.; Katsumata, K.-I.; Nakata, K.; Furudate, M.; et al. Photocatalytic, superhydrophilic, self-cleaning TiO$_2$ coating on cheap, light-weight, flexible polycarbonate substrates. *Appl. Surf. Sci.* **2018**, *458*, 917–923. [CrossRef]
17. Nundy, S.; Ghosh, A.; Mallick, T.K. Hydrophilic and Superhydrophilic Self-Cleaning Coatings by Morphologically Varying ZnO Microstructures for Photovoltaic and Glazing Applications. *ACS Omega* **2020**, *5*, 1033–1039. [CrossRef]
18. Esmeryan, K.D.; Ganeva, R.R.; Stamenov, G.S.; Chaushev, T.A. Superhydrophobic Soot Coated Quartz Crystal Microbalanc-es: A Novel Platform for Human Spermatozoa Quality Assessment. *Sensors* **2019**, *19*, 123. [CrossRef]
19. Esmeryan, K.D.; Chaushev, T.A. Complex characterization of human urine using super-nonwettable soot coated quartz crystal microbalance sensors. *Sens. Actuators A Phys.* **2021**, *317*, 112480. [CrossRef]
20. Esmeryan, K.D.; Lazarov, Y.; Stamenov, G.S.; Chaushev, T.A. When condensed matter physics meets biology: Does super-hydrophobicity benefiting the cryopreservation of human spermatozoa? *Cryobiology* **2020**, *92*, 263–266. [CrossRef]
21. Mallakpour, S.; Sirous, F.; Hussain, C.M. Green synthesis of nano-Al$_2$O$_3$, recent functionalization, and fabrication of syn-thetic or natural polymer nanocomposites: Various technological applications. *New J. Chem.* **2021**, *45*, 4885–4920. [CrossRef]

22. Radwan, A.B.; Abdullah, A.M.; Mohamed, A.M.A.; Al-Maadeed, M.A. New Electrospun Polystyrene/Al$_2$O$_3$ Nanocomposite Superhydrophobic Coatings; Synthesis, Characterization, and Application. *Coatings* **2018**, *8*, 65. [CrossRef]
23. Feng, L.; Zhang, H.; Mao, P.; Wang, Y.; Ge, Y. Superhydrophobic alumina surface based on stearic acid modification. *Appl. Surf. Sci.* **2011**, *257*, 3959–3963. [CrossRef]
24. Jagdheesh, R. Fabrication of a Superhydrophobic Al2O3 Surface Using Picosecond Laser Pulses. *Langmuir* **2014**, *30*, 12067–12073. [CrossRef] [PubMed]
25. Sutha, S.; Suresh, S.; Raj, B.; Ravi, K.R. Transparent alumina based superhydrophobic self–cleaning coatings for solar cell cover glass applications. *Sol. Energy Mater. Sol. Cells* **2017**, *165*, 128–137. [CrossRef]
26. Karapanagiotis, I.; Manoudis, P.N.; Savva, A.; Panayiotou, C. Superhydrophobic polymer-particle composite films produced using various particle sizes. *Surf. Interface Anal.* **2012**, *44*, 870–875. [CrossRef]
27. Richard, E.; Aruna, S.T.; Basu, B.J. Superhydrophobic surfaces fabricated by surface modification of alumina particles. *Appl. Surf. Sci.* **2012**, *258*, 10199–10204. [CrossRef]
28. Tie, L.; Li, J.; Liu, M.; Guo, Z.; Liang, Y.; Liu, W. Facile Fabrication of Superhydrophobic and Underwater Superoleophobic Coatings. *ACS Appl. Nano Mater.* **2018**, *1*, 4894–4899. [CrossRef]
29. Byun, H.R.; Ha, Y.G. Non-wetting superhydrophobic surface enabled by one-step spray coating using molecular self-assembled nanoparticles. *J. Nanosci. Nanotechnol.* **2017**, *17*, 5515–5519. [CrossRef]
30. Kim, I.-S.; Cho, M.-Y.; Jeong, Y.; Shin, Y.-C.; Lee, D.-W.; Park, C.; Koo, S.-M.; Shin, W.H.; Yang, W.-J.; Park, Y.; et al. Aerosol-deposited Al2O3/PTFE hydrophobic coatings with adjustable transparency. *J. Am. Ceram. Soc.* **2021**, *104*, 1716–1725. [CrossRef]
31. Na, M.J.; Yang, H.; Jung, H.J.; Park, S.D. Robust hydrophobic surface driven by Al$_2$O$_3$/glass composite coatings. *Surf. Coat. Technol.* **2019**, *372*, 134–139. [CrossRef]
32. Jeon, Y.; Nagappan, S.; Li, X.-H.; Lee, J.-H.; Shi, L.; Yuan, S.; Lee, W.-K.; Ha, C.-S. Highly Transparent, Robust Hydrophobic, and Amphiphilic Organic–Inorganic Hybrid Coatings for Antifogging and Antibacterial Applications. *ACS Appl. Mater. Interfaces* **2021**, *13*, 6615–6630. [CrossRef] [PubMed]
33. Naskar, M.K. Soft Solution Processing for the Synthesis of Alumina Nanoparticles in the Presence of Glucose. *J. Am. Ceram. Soc.* **2010**, *93*, 1260–1263. [CrossRef]
34. Xue, G.; Huang, X.; Zhao, N.; Xiao, F.; Wei, W. Hollow Al$_2$O$_3$ spheres prepared by a simple and tunable hydrothermal method. *RSC Adv.* **2015**, *5*, 13385–13391. [CrossRef]
35. Ghosh, S.; Dalapati, R.; Naskar, M.K. Understanding the role of tetramethyl urea for the synthesis of mesoporous alumina. *J. Asian Ceram. Soc.* **2014**, *2*, 380–386. [CrossRef]
36. Latthe, S.S.; Demirel, A.L. Polystyrene/octadecyltrichlorosilane superhydrophobic coatings with hierarchical morphology. *Polym. Chem.* **2012**, *4*, 246–249. [CrossRef]
37. Pawar, P.G.; Xing, R.; Kambale, R.C.; Kumar, A.M.; Liu, S.; Latthe, S.S. Polystyrene assisted superhydrophobic silica coatings with surface protection and self-cleaning approach. *Prog. Org. Coat.* **2017**, *105*, 235–244. [CrossRef]
38. Xue, C.-H.; Zhang, Z.-D.; Zhang, J.; Jia, S.-T. Lasting and self-healing superhydrophobic surfaces by coating of polysty-rene/SiO2 nanoparticles and polydimethylsiloxane. *J. Mater. Chem. A* **2014**, *2*, 15001–15007. [CrossRef]
39. Nagappan, S.; Ha, C.S. In-situ addition of graphene oxide for improving the thermal stability of superhydrophobic hybrid materials. *Polymer* **2017**, *116*, 412–422. [CrossRef]
40. Zhang, J.; Zhang, L.; Gong, X. Large-Scale Spraying Fabrication of Robust Fluorine-Free Superhydrophobic Coatings Based on Dual-Sized Silica Particles for Effective Antipollution and Strong Buoyancy. *Langmuir* **2021**, *37*, 6042–6051. [CrossRef]
41. Zhang, C.; Huo, R.; Wang, X.; Zhang, J.; Cheng, J.; Shi, L. In-situ encapsulation of flaky aluminum pigment with poly(methylhydrosiloxane) anti-corrosion film for high-performance waterborne coatings. *J. Ind. Eng. Chem.* **2020**, *89*, 239–249. [CrossRef]
42. Fang, C.; Pu, M.; Zhou, X.; Lei, W.; Pei, L.; Wang, C. Facile preparation of hydrophobic aluminum oxide film via sol-gel method. *Front. Chem.* **2018**, *6*, 308–313. [CrossRef] [PubMed]
43. Rodič, P.; Kapun, B.; Panjan, M.; Milošev, I. Easy and fast fabrication of self-cleaning and anti-icing perfluoroalkyl silane film on aluminium. *Coatings* **2020**, *10*, 234. [CrossRef]
44. Tong, W.; Xiong, D.; Wang, N.; Wu, Z.; Zhou, H. Mechanically robust superhydrophobic coating for aeronautical composite against ice accretion and ice adhesion. *Compos. Part B Eng.* **2019**, *176*, 107267. [CrossRef]
45. Cai, C.; Sang, N.; Teng, S.; Shen, Z.; Guo, J.; Zhao, X.; Guo, Z. Superhydrophobic surface fabricated by spraying hydrophobic R974 nanoparticles and the drag reduction in water. *Surf. Coat. Technol.* **2016**, *307*, 366–373. [CrossRef]
46. Hill, D.; Barron, A.; Alexander, S. Controlling the wettability of plastic by thermally embedding coated aluminium oxide nanoparticles into the surface. *J. Colloid Interface Sci.* **2020**, *567*, 45–53. [CrossRef]
47. Shah, S.; Zulfiqar, U.; Hussain, S.; Ahmad, I.; Hussain, I.; Subhani, T. A durable superhydrophobic coating for the protection of wood materials. *Mater. Lett.* **2017**, *203*, 17. [CrossRef]

Article

Robust Superhydrophobic and Repellent Coatings Based on Micro/Nano SiO₂ and Fluorinated Epoxy

Xiaoye Huang and Ruobing Yu *

School of Materials Science and Engineering, East China University of Science and Technology, No. 130 Meilong Road, Lingyun District, Shanghai 200237, China; y30180417@mail.ecust.edu.cn
* Correspondence: rbyu@ecust.edu.cn; Tel.: +86-133-3198-2086

Abstract: Superhydrophobic surfaces possess low mechanical strength, and can be easily contaminated by fluids with low surface tension, such as oil; this hinders their practical applications. In this study, fluorinated epoxy was prepared through the thiol-ene click reaction at first. The superhydrophobic surface with high oil-repellency was prepared by the addition of unmodified nano-SiO$_2$ and micron-SiO$_2$ to the fluorinated epoxy. The effect of the ratio of micro- and nano-silica particles on the morphology and wettability of the coating was investigated. It was shown that a re-entrant structure appears and FEP-S coating has good liquid repellency when the amounts of nano-SiO$_2$ and micro-SiO$_2$ are equal. The contact angles of the FEP-S coating (coating with the best liquid repellent performance) for water, glycerol, ethylene glycol, and diiodomethane were 158.6° ± 1.1°, 152.4° ± 0.9°, 153.4° ± 1.3°, and 140.7° ± 0.9°, respectively. In addition, the superhydrophobic coatings possess excellent mechanical and chemical durability, excellent performance in self-cleaning, corrosion resistance, and anti-icing properties. The preparation method of superhydrophobic coating is relatively simple; therefore, it has a wide range of applications and can also be applied to various substrates.

Keywords: superhydrophobic; oleophobic; click chemistry; silica; fluorinated epoxy

Citation: Huang, X.; Yu, R. Robust Superhydrophobic and Repellent Coatings Based on Micro/Nano SiO$_2$ and Fluorinated Epoxy. *Coatings* **2021**, *11*, 663. https://doi.org/10.3390/coatings11060663

Academic Editors: N. P. Prorokova and Ioannis Karapanagiotis

Received: 10 April 2021
Accepted: 28 May 2021
Published: 31 May 2021

Publisher's Note: MDPI stays neutral with regard to jurisdictional claims in published maps and institutional affiliations.

1. Introduction

The superhydrophobic surface is a surface where the water contact angle is greater than 150° and the contact angle hysteresis is lower than 10°. Superhydrophobic surfaces have wide applications in self-cleaning [1], antifouling [2], anti-icing [3,4], anti-corrosion [5,6], oil transfer, and oil-water separation [7,8]. Extensive literature is available related to the preparation of superhydrophobic surfaces. However, in practical applications, the superhydrophobic surfaces inevitably encounter some problems. After contact with organic liquids with low surface tension, and even fingerprints, the surface will lose their superhydrophobicity. Therefore, superhydrophobic surfaces which can repel both water and oil have more practical applications as compared with ordinary superhydrophobic surfaces.

However, superoleophobic surfaces have low surface energy and their manufacturing is difficult than that of ordinary superhydrophobic surfaces. In addition, superoleophobic surfaces possess more delicate structures, such as re-entrant structures. The superoleophobic surfaces can repel fluids with low surface tension when the droplet is in the Cassie-Baxter state. Since Tuteja et al. [9] developed a superoleophobic surface by introducing re-entrant features in 2007, several researchers have investigated the superoleophobic surfaces by designing a hierarchical structure similar to a re-entrant structure (e.g., hanging structure [10,11], inverted trapezoid structure [12,13], mushroom shape [8,14], flower shape [15–17], bowstring shape [18], nano-filament [19,20] and candle soot [21]). However, re-entrant structures are more delicate than simple hierarchical structures, which are difficult to manufacture and require complex techniques and expensive equipment, for example, electrospinning [22,23], lithography [24,25], templating [21,26], laser ablation [27–29], anodic oxidation [30,31], vapor deposition [2,32], plasma etching [33–35], and other combina-

13

tional approaches. Owing to the structural requirements and limitations of the professional equipment of the superoleophobic coating, it is more practical to fabricate a superhydrophobic and oleophobic coating by a simple method. In fact, the spray-coating or drop-coating method, which is based on an adhesive (e.g., Epoxy [36,37], polyurethane [38], inorganic adhesive [39] and 3M glue [14]) and nanoparticles (e.g., SiO_2 [40], TiO_2 [14] and ZnO [17]), is simple and economical for the fabrication of superhydrophobic and oleophobic coatings.

Xiong et al. [40] synthesized two block copolymers by anionic polymerization, poly [3-(triisopropyloxysilyl)propyl methacrylate]-block-poly(perfluorooctylethyl methacrylate) (PIPSMA-b-PFOEMA) and poly(tert-butyl acrylate) (PIPSMA-b-PtBA), which were grafted onto SiO_2 through the Stober method in the presence of HCl. The as-prepared bi-functional silica solution was drop-casted onto epoxy glue (first coated on a glass slide). The coating can repel water and oil with good adhesive strength. However, the synthesis of two block copolymers and bi-functional silica particles is complex.

Wang et al. [36] added nano-silica and carbon nanotubes to epoxy resin (EP)/modified poly (vinylidene fluoride) (MPVDF)/fluorinated ethylene propylene (FEP) composite, and obtained a superamphiphobic coating with a high wear life and corrosion resistance.

Su et al. [41] first obtained modified micron- and nano-silica by grafting epoxy resin, and then sprayed these on the glass slides in turn with a spray gun. The coating was superhydrophobic. The preparation and curing time of the coating were long.

Li et al. [37] prepared fluorinated nano-SiO_2 particles by the sol-gel method and used these as a sheath with a core of epoxy solution to prepare a superamphiphobic coating by a coaxial electrospray method. However, the wear resistance of the coating was low.

Zhang et al. [42] applied bisphenol A diglycidyl ether (BADGE) epoxy resin and unmodified multi-walled carbon nanotubes (MWCNTs) to establish a superhydrophobic coating, which exhibited good mechanical durability and anti-icing performance.

Xiu et al. [43] mixed silica particles (100 nm) and bisphenol A diglycidyl ether, hexahydro-4-methylphthalic anhydride, and imidazole in toluene and coated the dispersion on a glass slide, which was then cured for 4 h at 150 °C. The epoxy was etched away and silica nanoparticles were exposed on the surface by plasma etching. The coating was then dipped in perfluorinated octyl trichlorosilane (PFOS) to form a superhydrophobic surface.

Han et al. [44] prepared raspberry-like hollow SnO_2 nanoparticles by a hydrothermal method and modified it with 1H, 1H, 2H, 2H-Perfluorodecyltriethoxysilane (FAS17). The modified SnO_2 and SiO_2 were added to the epoxy to obtain a robust superamphiphobic coating with good stability.

Peng et al. [45] first sprayed epoxy resin on the matrix as an adhesive, and then zinc oxide and silica particle solution were coated on the surface to construct a rough surface. The surface was wear-resistant and superamphiphobic.

Aslanidou et al. [46] have prepared a water soluble siloxane emulsion enriched with silica nanoparticles (7 nm) and sprayed it on the surface of marble and sandstone. The coating is superhydrophobicity and superoleophobicity. It was shown that when the nanoparticle concentration is 2% w/w, the coating has best superhydrophobicity and superoleophobicity.

From the aforementioned discussion, there are two procedures to prepare the coating by "glue + particles". One is to spray the mixture of particles and glue, which can improve the adhesive strength of the coating; however, it is difficult to achieve oleophobic properties because the particles are covered with the glue. The other method is to coat the particle solution on the surface of the glue. The exposed particles can easily fall owing to the low adhesive strength. In this study, we aimed to prepare a superhydrophobic and oleophobic coating using a mixture of particles and glue. To achieve oleophobic performance, a novel fluorinated epoxy, which can provide both good adhesion and strong liquid repellency, was employed as the glue. The synthesis of fluorinated epoxy containing epoxy groups and long perfluoroalkyl chains was based on the modified Zhang's method [1,47,48]. This novel fluorinated epoxy was prepared by introducing 1H, 1H, 2H, 2H-perfluorodecyl acrylate (PFDA), and glycidyl methacrylate (GMA) into pentaerythritol tetra (3-mercaptopropionate) (PETMP) via a simple and fast thiol-ene click reaction. By adding micro- and nano-silica particles to fluorinated epoxy and adjusting their proportions, we obtained a superhydrophobic and oleophobic coating with different micro/nanostructures and different liquid repellent properties.

2. Materials and Methods

2.1. Materials

Acetone (99.5%) was purchased from General-Reagent; PFDA was obtained from Adamas-beta. GMA and PETMP were supplied by Aladdin; 2-methylimidazole (GC), 2, 2-bimethoxy-2-phenylacetophenone (DMPA), micron silicon dioxide (W-180, with an average particle size of 2 μm), and nano silicon dioxide (S861577, with an average particle size of 20 nm) were sourced from Macklin.

2.2. Preparation of Fluorinated Epoxy

Fluorinated epoxy was fabricated by a modified method of Zhang et al. [1,47,48]. We applied thiol-ene click chemistry reaction (Figure 1) to prepare fluorinated epoxy as follows: 0.3 mol GMA, 0.1 mol PETMP and 0.1 mol PFDA were mixed in acetone, and then the mixture was treated by ultrasonic for 10 min before adding 0.001 mol DMPA. The solution was exposed to ultraviolet radiation (365 nm, 300 W) for 1 h at 25 °C.

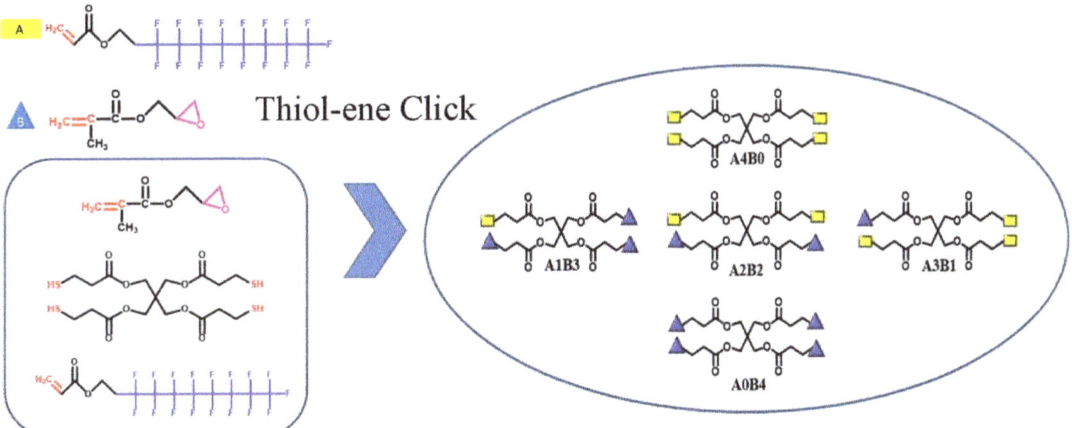

Figure 1. Scheme for preparation of fluorinated epoxy via thiol-ene click method.

2.3. Preparation of Micro/Nano-Silica Coatings by Spray-Coating Approach

The procedure for the preparation of the coating is shown in Figure 2. First, nano- and micron-SiO_2 particles with mass ratios of 0:1, 1:0, 2:1, 1:2, and 1:1 were ultrasonically dispersed in 3 g of fluorinated epoxy (50 wt% in acetone solution) for 20 min. Subsequently, five as-prepared solutions were added to 15 mg GC and magnetically ultrasonicated for 20 min at 25 °C. Solutions with five different ratios were obtained (0:1, 1:0, 2:1, 1:2, and 1:1) and marked as S1, S2, S3, S4, and S5, respectively.

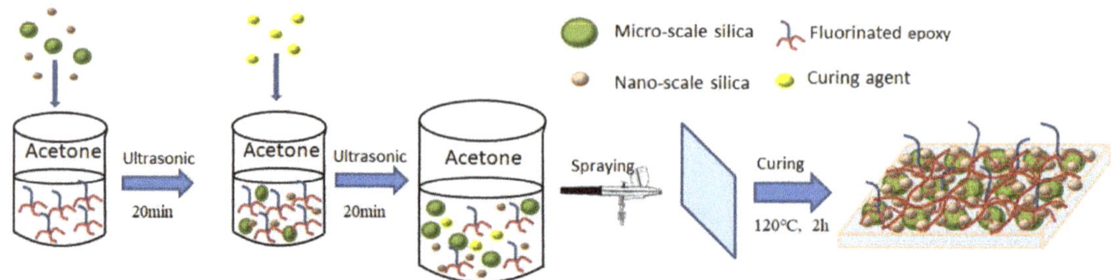

Figure 2. Scheme for preparation of coating by spraying.

Before spray coating, all glasses and aluminum foil were placed in deionized water and ethanol for ultrasonic cleaning and then blow-dried in air. Subsequently, the as-prepared solution was sprayed onto the slides or aluminum foil by a spray gun with a nozzle diameter of 0.5 mm and 0.3 MPa compressed air (Youlun S-131, Master airbrush). The operating air was controlled by an airbrush compressor (Lotus T-300K), and the distance between the airbrush and the substrate was set at approximately 15 cm. The coated slides were placed on a heating platform at 80 °C for 5 min to remove the solute and then placed in an oven at 120 °C for 2 h.

We named the coating with the best liquid repellent performance as FEP-S (with solution S5). In order to understand the comprehensive property of FEP-S, the surface chemical composition, mechanical and chemical stability and self-cleaning properties and anti-icing performance were also evaluated.

2.4. Analysis and Testing Methods

- Abrasion test

Sandpaper (600 grid) and a standard weight (100 g) were used to abrade FEP-S coating at a pressure of 1 kPa. The sandpaper was placed on a horizontal surface and the coating was placed on the sandpaper. A weight was placed on the back of the coating, and the coating was slid along the ruler at 1 cm/s on the sandpaper surface, 10 cm to the right, and 10 cm to the left for a cycle test. We measured the CA of FEP-S coating for the four liquids after every 10 cycles.

- Cross cut adhesion test

With a cross-hatch cutter (BEVS2202, 2 mm), as per the Chinese National Standard (GB/T 9286-1998), a lattice pattern was cut with equidistant spacing on the coating surface, and commercial cellophane tape was applied over the lattice for 5 min and then peeled off. The adhesion level of the coating, as obtained from GB/T 9286-1998, is presented in Table 1.

- Chemical stability test

Water droplets with different pH values ranging from 1 to 14 were placed on the surface of the superhydrophobic coating to test the chemical stability.

The chemical stability of the FEP-S coating was further tested by immersing the samples in a strong acid (HCl, pH = 1) or basic solution (NaOH, pH = 14) at 25 °C for 1 h. The sample was then rinsed with distilled water for 3 min, and oven-dried at 80 °C for 10 min. Subsequently, the WCAs of the samples were recorded.

Table 1. Adhesion level of cross-cut method.

Adhesion Level	Description	Percent (%) Area Removed
0	The edges of the square are completely smooth, none of the squares of the lattice are detached	0%
1	Small flakes of the coating are detached at intersections, and less than 5% of the area is affected	<5%
2	Small flakes of the coating are detached along edges and at the intersections of cuts	5–15%
3	The coating has flaked along the edges and on parts of the squares	15–35%
4	The coating has flaked along the edges of the cuts in large ribbons and whole squares have detached	35–65%
5	Severe flaking and detachment across entire square	>65%

- Self-cleaning test

A self-cleaning test was conducted using carbon black powder and $CuSO_4$ powder as contaminants on the surface of the FEP-S coating. Water droplets were dripped onto the coated glass through a disposable dropper.

The repellent properties of the FEP-S coating were tested by immersing the samples in common liquids that are used in daily life (e.g., water, coffee, cola, juice, tea, and soybean milk) for 5 min, and then taken out to observe the surface.

- Anti-icing test

The ice delay property of the FEP-S coating was tested by dropping water (~0.05 mL) on the original glass slide and the coated glass slide, and these slides were placed in a freezer at $-18\ ^\circ$C. The slides were taken out after every 30 s to observe the state of the water droplets.

Dynamic anti-icing of the coating was also performed in the freezer at $-18\ ^\circ$C. Bare glass and coated glass slides were placed in the freezer for 24 h, and a few drops of water at $0\ ^\circ$C were dropped on the coating to observe the state of droplets falling on the coating.

- Other characterization

The FEP-S coating used in the SEM, EDS and XPS tests is based on an aluminum foil substrate, and the other characterizations are based on a glass slide substrate. The morphology (SEM photograph) and the element distribution of the coating was studied by field emission scanning electron microscopy (FE-SEM, S-4800, Hitachi, Japan) equipped with energy dispersive spectroscopy (EDS mapping). The sample was attached to a conductive adhesive and sprayed with gold. The acceleration voltage of the instrument was 10–20 kV.

The content and distribution of the surface elements of FEP-S were characterized by X-ray photoelectron spectroscopy (XPS, Thermo Scientific Escalab 250Xi, Massachusetts, MA, USA). Before XPS test, FEP-S coating was placed in a vacuum oven at $80\ ^\circ$C for 12 h. The contact angles (CAs) of the different liquids on the coating were measured using the contact angle tester (JC 2000D2, Zhongchen, China) at $25\ ^\circ$C. All CA values were determined by averaging the values at five different points on each sample surface.

The sliding angles (SAs) were measured with a contact angle system (Kruss DSA 100S, Hamburg, Germany).

IR data of fluorinated epoxy and a mixture of PFDA, GMA and PETMP were obtained by Fourier transform infrared spectrometer (Nicolet 6700, Thermo Fisher, Waltham, MA, USA) operating over the frequency range of 4000–500 cm^{-1}. The samples were prepared via potassium bromide tableting.

Proton nuclear magnetic resonance (1H-NMR) spectra were obtained using a III-400 NMR spectrometer (Bruker Avance, Zurich, Switzerland), and deuterated chloroform was used as the solvent. TMS was used as the internal standard.

3. Results and Discussion

3.1. Preparation and Characterization of Fluorinated Epoxy

Fluorinated epoxy was manufactured by a thiol-ene click reaction as proposed by Zhang [1,47,48]. The radical reaction is rapid and has no byproducts.

The chemical structure of the fluorinated epoxy was characterized by Fourier transform infrared spectroscopy (FT-IR) and proton nuclear magnetic resonance (1H-NMR). The comparison of typical IR spectra of the reactants (mixture of PETMP, GMA, and PFDA) and products (fluorinated epoxy) are presented in Figure 3. Absorption peaks of epoxy groups could be observed in both reactants and products, with wavelengths of 937 cm^{-1} and 906 cm^{-1}, respectively. The epoxy absorption peak was introduced into fluorinated epoxy by GMA, which can be cured and cross-linked under the action of a curing agent (GC) to achieve high bonding strength with the substrate. The absorption peaks at wavelengths of 1730 cm^{-1} for the reactants and products were the stretching vibration peaks of C=O, the absorption peak was due to the ester group present in the three reactants, and did not disappear after the clicking reaction. The absorption peaks of -CF$_2$ also presented in reactant and product, at wavelengths of 1148 cm^{-1} and 1204 cm^{-1} respectively. Compared with the reactants, the peak of C=C at 1636 cm^{-1} and the peak of -SH at 2585 cm^{-1} disappeared in the product owing to completion of reaction of -SH with C=C to form a -CS bond, which also led to an increase in the -CH peak around 2918 cm^{-1} in the product.

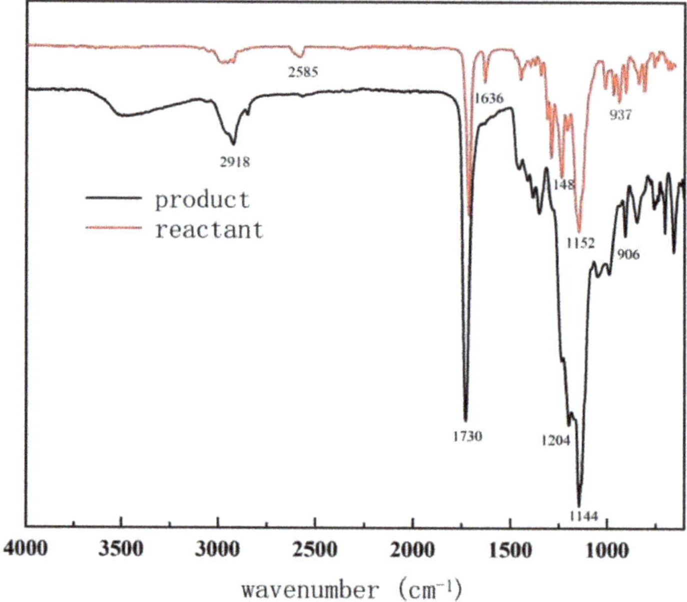

Figure 3. FT-IR spectra of the reactants (mixture of PFDA, GMA and PETMP) and product (fluorinated epoxy).

Figure 4 shows the 1H-NMR spectra of the reactants (tmixture of PETMP, GMA, and PFDA) and the products (fluorinated epoxy).

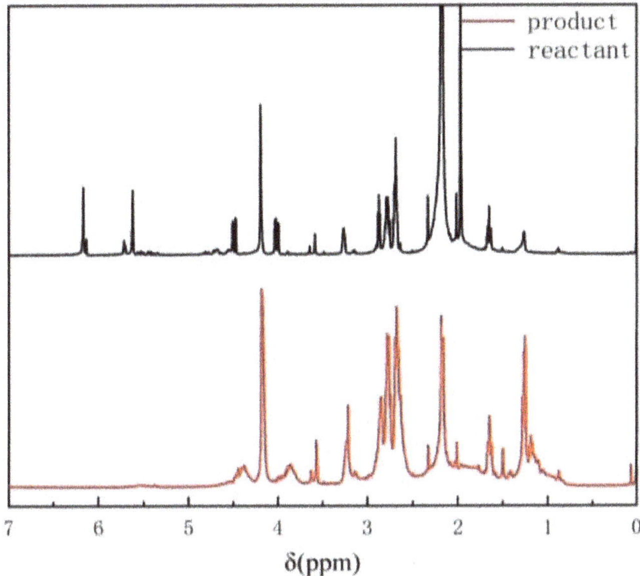

Figure 4. 1H-NMR spectra of the reactants (mixture of PFDA, GMA and PETMP) and product (fluorinated epoxy).

Vinyl terminal signals in GMA and PFDA were observed at 5.7–6.4 ppm in the 1 H-NMR spectra of the reactants. After the thiol-ene click reaction, the signal at 5.7–6.4 ppm disappeared, and the peak intensity of the methylene proton was enhanced at 2.6–2.8 ppm. In addition, the signal quantification of the methyl proton observed at 1.9 ppm adjacent to the GMA terminal double bond shifted to 1.2 ppm [47] owing to the change in chemical state of the methyl proton. The aforementioned observations prove that PETMP was successfully modified by GMA and PFDA to form the expected structure.

Although the molar ratio of GMA, PETMP and PFDA was 3:1:1, the four -SH groups on PETMP had the same probability of reacting with GMA and PFDA. Therefore, it is assumed that the case of four SH groups on PETMP reacting with GMA was C_4^0, the case of four SH groups on PETMP reacting with PFDA was C_4^4, the case of one SH group on PETMP reacting with GMA and three SH reacting with PFDA was C_4^1, the case of one SH group on PETMP reacting with PFDA and three SH reacting with GMA was C_4^3, the case of two SH group reacting with PFDA and two SH reacting with GMA was C_4^2. Thus there were $C_4^0 + C_4^1 + C_4^2 + C_4^3 + C_4^4$ cases in total. Therefore, the branched structure of fluorinated epoxy consisted of A2B2 (A is the group of PFDA in the fluorinated epoxy, and B is the group of GMA in the fluorinated epoxy), A1B3, A3B1, A0B4, and A4B0 types that yielded 37.5%, 25.0%, 25.0%, 6.25%, and 6.25%, respectively, as per the mathematical calculations. The content of A2B2, A1B3, A3B1, A0B4 and A4B0 were calculated as follows.

$$C_4^2 / \left(C_4^0 + C_4^1 + C_4^2 + C_4^3 + C_4^4 \right) = 37.5\%, \ (A2B2)$$
$$C_4^1 / \left(C_4^0 + C_4^1 + C_4^2 + C_4^3 + C_4^4 \right) = 25\%, \ (A1B3)$$
$$C_4^3 / \left(C_4^0 + C_4^1 + C_4^2 + C_4^3 + C_4^4 \right) = 25\%, \ (A3B1) \tag{1}$$
$$C_4^0 / \left(C_4^0 + C_4^1 + C_4^2 + C_4^3 + C_4^4 \right) = 6.25\%, \ (A0B4)$$
$$C_4^4 / \left(C_4^0 + C_4^1 + C_4^2 + C_4^3 + C_4^4 \right) = 6.25\%, \ (A4B0)$$

As the molecule synthesized by the thiol-ene click reaction of PETMP and GMA contains both epoxy groups and long perfluoroalkyl chains, it had significant adhesion strength with the matrix, and also made the coating itself hydrophobic (WCA was approximately $105.3 \pm 0.7°$).

3.2. Wetting Behavior and Mechanical Durability of the Coating

As the epoxy groups in fluorinated epoxies can react with Si–OH on silica and GC, the coating had a good bonding strength with the substrate and high mechanical durability. The composite system of micro-SiO_2 and nano-SiO_2 particles in the coating can improved the roughness of the surface, and the appropriate proportion of micro and nanoparticles can form a special re-entrant structure, which was the key for the establishment of a superhydrophobic and oleophobic coating.

The wetting behavior of the coatings were evaluated by measuring the CAs for both low- and high-surface-tension liquids, including water, glycerin, glycol, and di-iodomethane. The results are shown in Table 2 and Figure 5g. When the total amount of fixed silica was 10% of the mass of fluorinated epoxy, the mass ratio of nano- and micron-silica showed a significant effect on the repellant properties for different liquids. By the addition of micron silica into the fluorinated epoxy, the liquid repellency of the coating was poor, and WCA of coating was only $123.4 \pm 2.3°$. With an increase of nano silica content in the silica particles, the CA of the coating for four liquids increased significantly. That is because the addition of nano particles could change the micro/nanostructure of the coating. When the amounts of nano-SiO_2 and micron-SiO_2 were equal (5 wt% nano-silica and 5 wt% micro-silica of fluorinated epoxy), the CA of the coating for four liquids reached the maximum value, and we named this coating with the best liquid repellent performance as FEP-S. Figure 5a–d shows the state of different liquids on the FEP-S coating, including water droplet (72.8 mN·m^{-1}), glycerol (64.0 mN·m^{-1}), ethylene glycol (47.7 mN·m^{-1}) and diiodomethane (50.8 mN·m^{-1}). The liquid droplets could easily roll on the sloping surface. The SAs of the different liquids are listed in Table 3. With further increase in nano-SiO_2, agglomeration of nano-SiO_2 appeared, indicating that the nano-SiO_2 could not be uniformly dispersed on the surface of micron-SiO_2 to form a re-entrant structure, and the CA for four liquids decreased at the same time. In the presence of only nano-SiO_2, the liquid repellent property of the coating surface was similar to that of only micron-SiO_2, and the superhydrophobic effect was not achieved. In Section 3.3, we will further discuss the effect of the proportion of micro/nano particles on the structure and liquid repellency of the coating.

Table 2. CAs for different liquids of fluorinated epoxy coatings with different proportions of micro/nano-silica.

CAs (°)	Water	Glycerol	Glycol	Diiodomethane
micro:nano = 1:0	123.4 ± 2.3	105.8 ± 1.4	127.3 ± 1.1	113.6 ± 1.3
micro:nano = 2:1	135.3 ± 0.8	110.4 ± 1.3	128.6 ± 1.5	118.5 ± 2.1
micro:nano = 1:1	158.6 ± 1.1	140.7 ± 0.9	152.4 ± 0.9	153.4 ± 1.3
micro:nano = 1:2	125.2 ± 1.5	120.5 ± 1.5	132.7 ± 1.4	120.8 ± 1.8
micro:nano = 0:1	120.7 ± 2.7	114.4 ± 1.8	118.5 ± 1.7	108.5 ± 2.4

Table 3. SAs for different liquids of FEP-S coatings (proportions of micro/nano-silica is 1:1).

SAs (°)	Water	Glycerol	Glycol	Diiodomethane
FEP-S	8.6 ± 1.1	14.6 ± 1.2	18.4 ± 1.5	17.5 ± 1.4

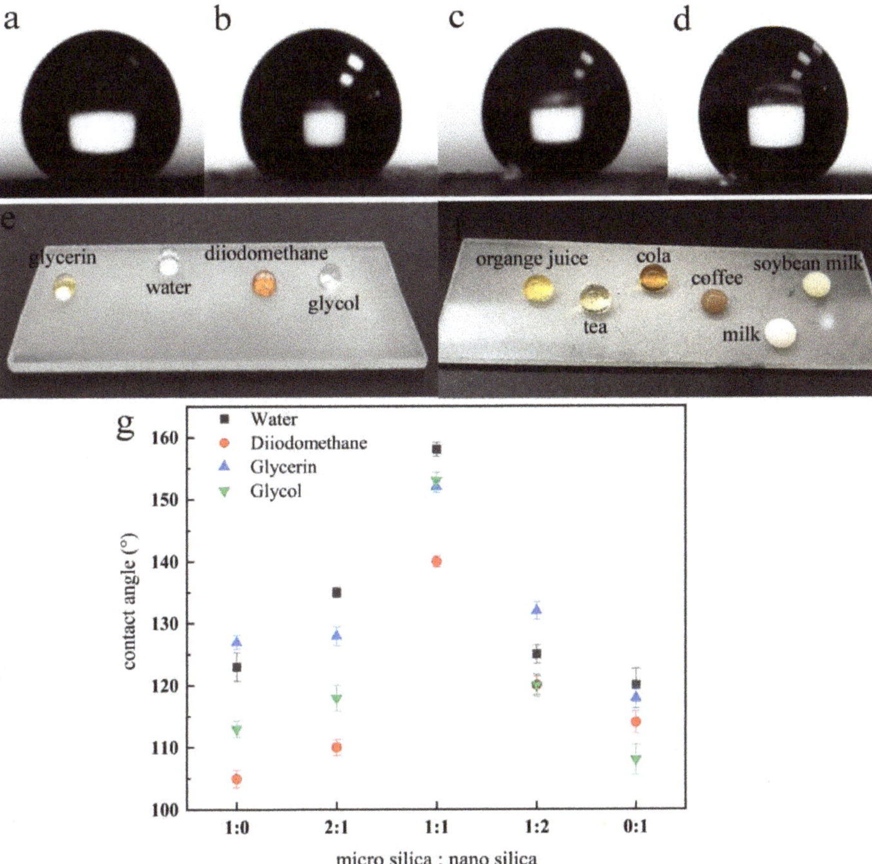

Figure 5. CAs of FEP-S slide for (**a**) water (**b**) diiodomethane (**c**) glycerol and (**d**) ethylene glycol (**e**,**f**) droplets of different liquids on FEP-S slide (**g**) CA for different liquids of coatings with different proportions of micro/nano-silica.

Durability is an important factor affecting the application of superhydrophobic materials. In particular, mechanical friction will lead to the damage of micro/nanostructures on the surface, and even slight scraping will cause function failure and loss of superhydrophobicity. Various methods are used to assess wear resistance, including sandpaper wear tests, sand/water impact, knife scraping, and tape stripping tests. In this study, the most extensive sandpaper wear test and a cross cut test were employed to evaluate the wear resistance and adhesion of the coating, respectively.

To investigate the effect of the proportion of micron- and nano-silicon on the adhesion of the coating, five coatings with different proportions of micron- and nano-silica (the mass ratio of silica particles was constant, and it was 10 wt% of fluorinated epoxy) was tested using a cross cut test. It was observed that the adhesion of all the five coatings was at level 0, indicating no falling-off of any grid or intersection, and the surface was perfect after cross cutting. In addition, when the silica particles was 10 wt% of the proportion of the fluorinated epoxy, silica particles could evenly disperse and cross-link with the fluorinated epoxy. When the fluorinated epoxy was completely cured in the presence of GC, the silica and fluorinated epoxy composite system exhibited a significant adhesion strength with the substrate.

The mass ratio of silica particles and fluorinated epoxy has an influence on the adhesion strength. In this study, the adhesive strength of the coating with different amounts of silica particles (including micro and nano silicon, and the ratio of nano-silica and micron silica is 1:1) was explored. Consequently, when silica particles reached 10% of the mass of fluorinated epoxy (FEP-S coating), the adhesion strength of the coating was at grade 0 (the optimal level). When silica particles were 20% of the fluorinated epoxy, the adhesion strength of the coating was insignificant. Meanwhile, the bumps of silica on the coating surface could be easily taped off. The damaged area was 5% or less, and the adhesion was at level 1. When the amount of silica particles reached 30% of the mass of the fluorinated epoxy, the silica particles aggregated. The bumps of silica on the coating surface were evident, and the particles on the intersection of the grid edge were observed, which could be easily removed. The total damaged area was less than 15%, and its adhesion grade was 2.

Here, an FEP-S coating with the best liquid repellent performance was selected to test the wear resistance. The operational steps are shown in Figure 6a. A 600 mesh sandpaper was placed on a horizontal surface, and the coating was placed on the sandpaper to make the better contact of coating with the sandpaper. A 100 g weight was placed on the back of the coating, and the coating was slide along the ruler at 1 cm/s on the sandpaper surface, 10 cm toward the right and 10 cm toward the left for a cycle test. The CA of four liquids was measured after every 10 cycles, as shown in Figure 6c. The liquid repellent property of the coating to the four liquids decreased with an increase in the friction. This is because the micro/nanostructures on the surface of the coating were slightly destroyed after the sandpaper test, resulting in the loss of the rough structure. After 50 wear cycles, the coating still exhibited superhydrophobic ability, indicating that the WCA was still greater than 150.0°.

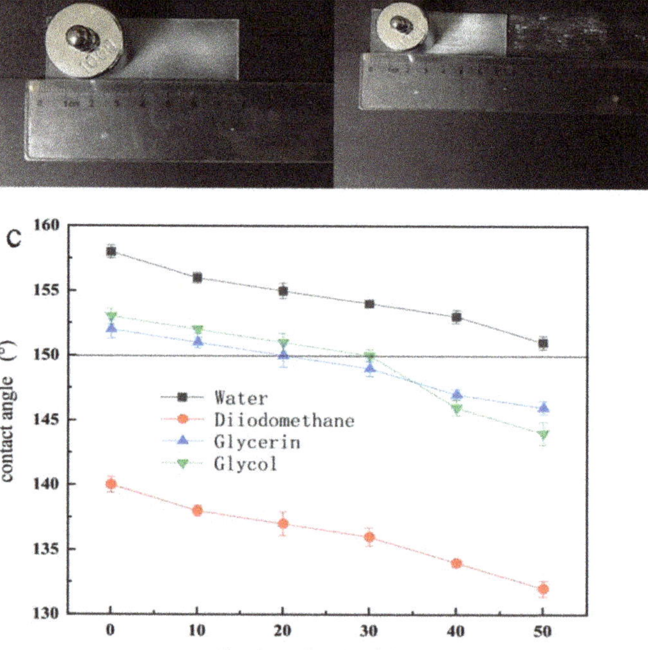

Figure 6. Image of the FEP-S coating (**a**) before and (**b**) after sandpaper abrasion at 10 m, (**c**) CA of four liquids changes of the FEP-S with sandpaper abrasion cycles.

As the thickness had a certain influence on the mechanical stability of the coating, especially on the abrasion resistance, the thickness of the FEP-S coating before and after sandpaper abrasion at 10 m was measured by SEM.

Figure 7a shows a cross-sectional SEM image of the FEP-S coating, and the thickness of FEP-S was approximately 110.7 µm, which was higher than the superhydrophobic or superoleophobic coatings (50–100 µm) prepared by the spray method in the literature [49,50]. From the SEM images (Figure 7b) after abrasion at a distance of 10 m, the thickness of FEP-S coating was down to 35.6 µm, and the substrate surface was fully covered by the superhydrophobic coating. The large thickness of the FEP-S coating was one of the reasons for its good mechanical properties.

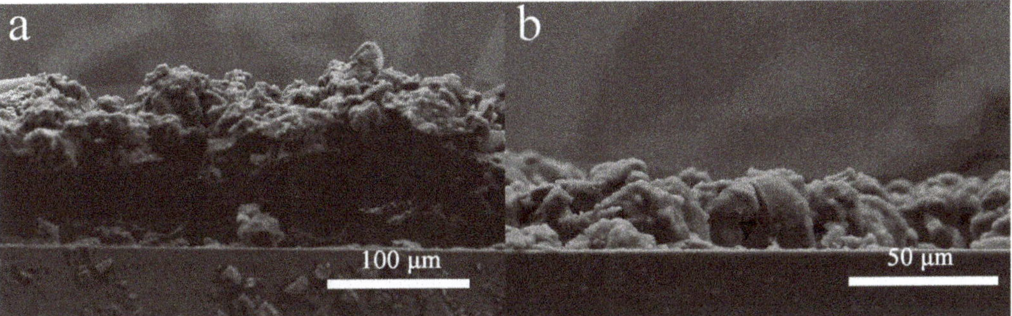

Figure 7. Cross-sectional SEM image of FEP-S coating (**a**) before and (**b**) after sandpaper abrasion at 10 m.

3.3. Surface Chemical Composition and Morphology

Figure 8(a$_1$–e$_1$) are the SEM images of S1–S5 coatings at 500× magnification, Figure 8(a$_2$–e$_2$) show the SEM images of S1–S5 coatings at 20,000× magnification, and Figure 8(f$_1$,f$_2$) show the SEM images of S5 at 35,000× and 60,000× magnifications. It can be observed that the particles on the five coatings were uniformly coated by fluorinated epoxy resin, and different micro/nanostructures are seen. Figure 8a shows that the surface of S1 coating (pure micron silica) was relatively flat, and nanoscale-bulges are not evident. From Figure 5g, the WCA of S1 coating was only 123.4°. Thus, the structure of S1 coating did not lead to the appearance of superhydrophobicity. In Figure 8b (pure nano-silica), the scale of the silica bump on the coating was approximately 2–30 µm, and the micronipples in the figure are marked by red circles. It can be seen that the degree of roughness of the coating was greater than that of pure micron silica. However, no nanoscale bulges were observed on the surface, because the agglomeration of nanoparticles coated with fluorinated epoxy forms a micron-scale structure, which had fewer nanoscale convex structures. As shown in Figure 8c (the ratio of nanoparticles to micron particles was 2:1), the surface morphology was similar to that in Figure 8b, but the size of micronipples was reduced and the distribution was more uniform, because the introduction of micron silica helped the dispersion of nano-silica. However, the introduction of a small amount of micron silica still failed to solve the problem of agglomeration of nano-silica particles, which is still unable to form nanoscale protrusion structures. When the ratio of nanoparticles to micron particles was 1:2, nano-silica particles were more evenly distributed on the micronipples and formed a nanoparticle bump (marked with a red circle in Figure 8d) at the same time. The results show that when the amount of micron silica was larger than that of nano-silica, the micro/nanostructure was formed on the surface, but the air gap structure that was formed could not retain the drop in the Cassie–Baxter state. Therefore, the liquid repellent performance of the coating at this ratio of micron/nano-silica was not the best. Figure 8e shows the SEM image of the coating surface (the ratio of nanoparticles to micron particles was 1:1); it is clear that the coating had a relatively uniform nanoscale roughness, and micron particles and aggregates also existed. In this case, the coating exhibited the best superhydrophobic and oleophobic properties. Figure 8f shows an enlarged image of S5 coating. Micronipples from nanoscale particle aggregates (diameters in the range of 0.5–1 μm) and nanoparticle bumps (diameters of ~50 nm) appeared, and we could see a large air gap between the micronipples. This structure was the same as the re-entrant structure, which allowed air to exist beneath the liquid droplet and causes the low-surface-tension droplet to remain Cassie–Baxter state on the surface. This forms the basis for repellency to water and low-surface-tension liquids, such as ethylene glycol.

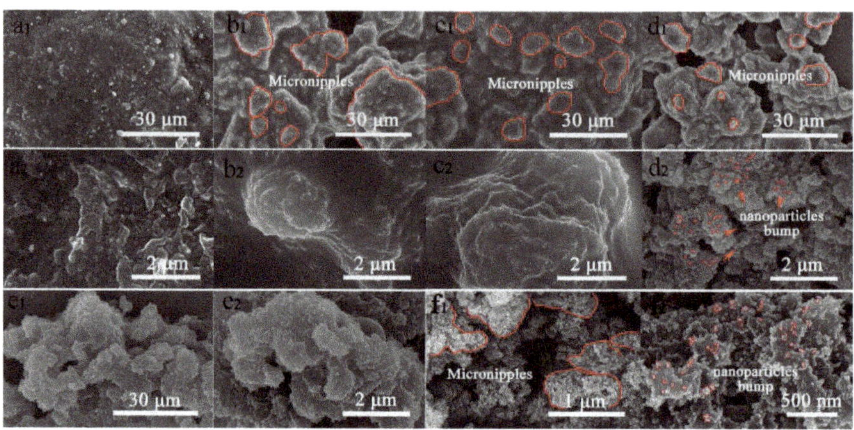

Figure 8. SEM image of fluorinated epoxy coating with different proportions of micro- and nano-silica (The mass ratio of nano-silica and micro-silica particles: (**a₁,a₂**) 0:1, (**b₁,b₂**) 1:0, (**c₁,c₂**) 2:1, (**d₁,d₂**) 1:2, (**e₁,e₂**) 1:1 and (**f₁,f₂**) 1:1 energy-dispersive X-ray spectroscopy (EDS) and X-ray photoelectron spectroscopy (XPS) were used to analyze the chemical composition of the FEP-S coating. C, O, F, and Si were detected in the EDS mapping shown in Figure 9; all elements were uniformly distributed.

Figure 9g shows the XPS spectrum in which some peaks corresponding to C1s (291.3 eV), O1s (532.2 eV), F1s (688.3 eV), and Si2p (101.8 eV) are presented. Figure 9e presents the high-resolution spectrum of C1s; the CF_2 (291.3 eV), CF_2-CH_2 (288.4 eV), C=O (288.9 eV), C-O/C-N and C-S (286.2 eV), and C-C/C-H (284.7 eV) peaks were detected. The absence of the peak of CF_3 in the high-resolution spectrum of C1s may be owing to the fact that CF_2 groups were mainly in the long perfluoroalkyl chains. Figure 9f shows the high-resolution spectrum of F1s, in which three fitting peaks could be assigned to CF_3 (689.3 eV), CF_2 (688.1 eV), and CF_2-CH_2 (686.9 eV), indicating the presence of a low surface energy -CF_3 group on the coating surface.

It can be seen from XPS that the amount of element F on the surface of the aluminum foil was only 11.9%, and it primarily existed in the form of CF_2 and CF_3 on the surface of the sample (located at 291.3 eV). This is because the long perfluoroalkyl chains on the PFDA molecule were grafted to the fluorinated epoxy molecule. The nano- and micron-silica were not treated with fluorosilane, so the fluorine content on the surface was low, which indicates that the properties of superhydrophobic and oleophobic of the coating was provided by the micro/nano rough structure on the surface and the low surface energy of fluorinated epoxy. Meanwhile, the Si content on the surface of the coating was only 2.64%, which means that the fluorinated epoxy covered most of the surface of the coating, so the coating exhibited good adhesion strength and wear resistance.

3.4. Chemical Stability

Corrosion resistance is an important property of superhydrophobic coating, which determines whether the coating can be applied in acidic or alkaline environment. Therefore, we characterized the chemical stability of the coating by two methods. Water droplets with pH values of 1–14 were placed on the surface of the FEP-S coating to test the chemical stability of the coating (as shown in Figure 10). Results show that droplets rolled on the coating without sticking. This indicates that the coating had strong repellent ability toward distilled water as well as a corrosive solution.

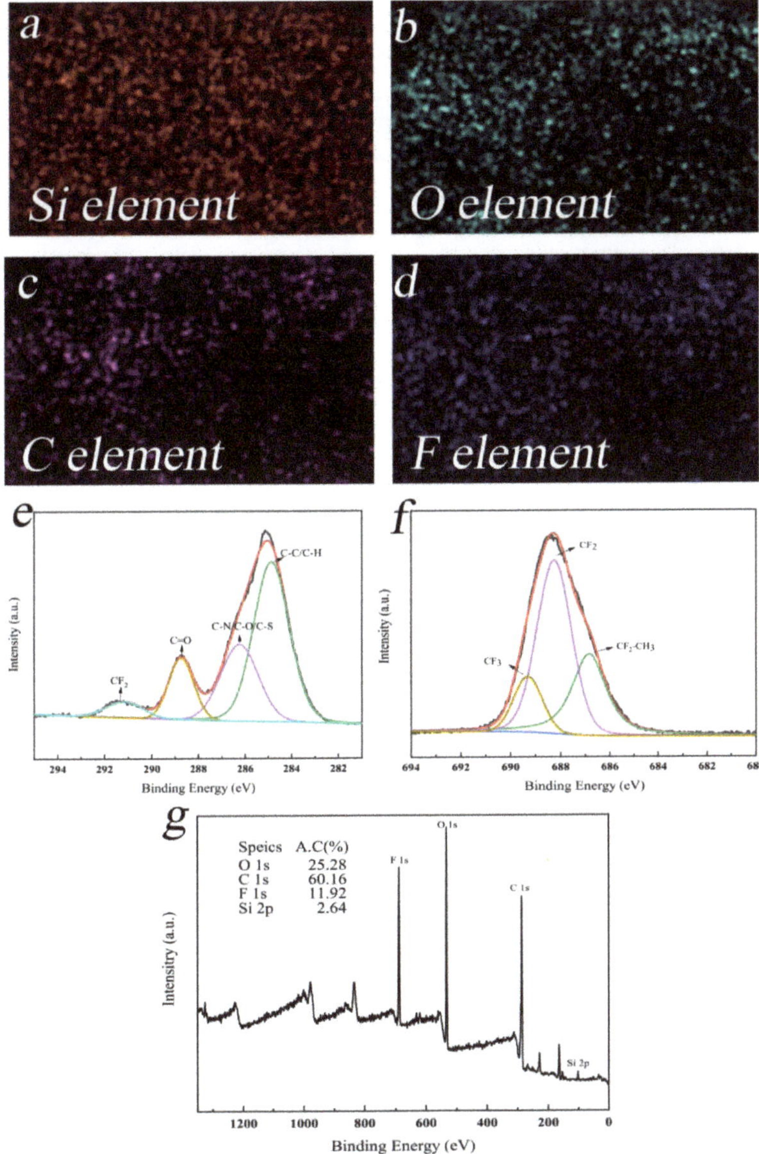

Figure 9. (**a–d**) EDS mapping, (**e**) high-resolution C1s spectra, (**f**) high-resolution F1s spectra and (**g**) XPS survey spectra of FEP-S coating.

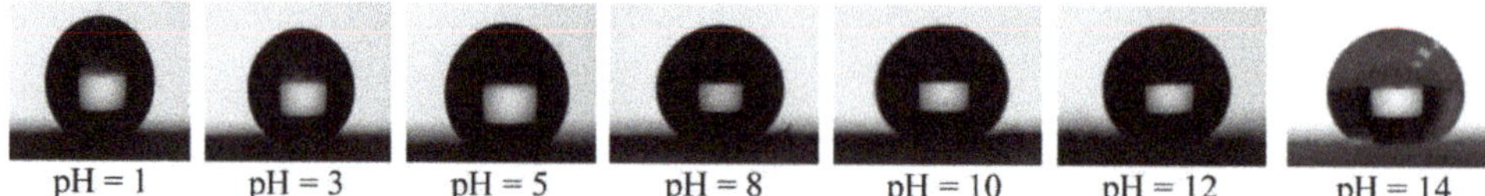

Figure 10. Shape of different pH solution droplets on FEP-S coating.

For further evaluation, chemical corrosion resistance of FEP-S in strong acidic and alkaline environments, the coating was placed in 1 mol/L hydrochloric acid and sodium hydroxide for 1 h, and then taken out, washed with distilled water for 3 min, and dried in an oven at 80 °C for 10 min. The CAs of FEP-S for water, glycerol, diiodimethane, and ethylene glycol are listed in Table 4.

Table 4. CA of FEP-S for different liquids after immersion in HCl and NaOH solution for 1 h.

CAs (°)	Water	Glycerol	Diiodomethane	Glycol
1M HCl	158.4 ± 0.8	152.5 ± 0.7	135.3 ± 1.1	147.5 ± 0.9
1M NaOH	157.6 ± 0.7	153.7 ± 0.6	136.4 ± 0.9	148.6 ± 1.2

As can be seen from Table 4, the FEP-S coating was still superhydrophobic after immersion in strong acid and alkali for 1 h. The WCA was approximately 158.0°, and the CA for glycerol is more than 150.0°. Meanwhile, the CAs for diiodide and ethylene glycol was still very high. The as-prepared FEP-S coating had excellent chemical stability and good acid and alkali resistance. This can be attributed to the long perfluoroalkyl chains on the FEP-S coating, silica, and the epoxy itself.

3.5. Self-Cleaning Properties

Lotus leaves have a natural superhydrophobic surface and good self-cleaning ability. The rolling of water droplets on the lotus leaves automatically carried away dust and kept the surface clean. An oil-free superhydrophobic surface can prevent the surface damage caused by oil to maintain self-cleaning ability for a long time. Figure 11 shows the repellent properties of the coating for common liquids, including orange juice, soy milk, black tea, coffee, cola, and milk. The FEP-S coating was immersed in these solutions for 5 min, and no liquid drops were found on the glass sheet, indicating that the coating was difficult to be polluted by common liquids in daily life.

Figure 11. Repellent properties of the FEP-S coatings for common liquids.

We used carbon black powder insoluble in water (which is harder to be cleaned than general dusts) and $CuSO_4$ powder soluble in water as pollutant to test the self-cleaning performance of the FEP-S coating. First, $CuSO_4$ powder and carbon black powder were evenly distributed over the surface as shown in Figure 12. Subsequently, water was dropped on the surface. It can be observed that pollutants easily rolled away from the surface along with the rolling of water droplets. The surface polluted by dust was cleaned in a short time, presenting the same clean surface state as before. Thus, in practical applications, the FEP-S coating could efficiently protect the substrate from dusts and other pollutants. These properties made the FEP-S coating able to be applied to more practical applications, such as kitchenware, exterior wall, oil pipeline, clothing etc.

Figure 12. Self-cleaning behavior of FEP-S coating using (**a–c**) $CuSO_4$ powder and (**d–f**) carbon black powder as contaminants.

3.6. Anti-Icing Performance

Ice often causes problems in transportation, such as power transmission and wind power generation. The reasonable construction of a superhydrophobic surface is promising for the application of anti-icing. In order to test the anti-icing performance of the FEP-S coating, a drop of water (~0.05 mL) was dropped on the original glass sheet and the glass sheet was sprayed with FEP-S coating, and the glass sheets were placed in a refrigerator at $-18\ °C$. The glass sheets were taken out every 30 s to observe the state of water droplets on the surface.

On the bare glass slide, the water drop began to freeze at ~15 s and was completely frozen at ~30 s (Figure 13). On the coated glass slide, the water drop began to freeze at 5 min and was completely frozen at 10 min. This means that the FEP-S coating could delay icing by ca. 10 min in extremely freezing weather. This may be due to the small solid–liquid contact area of the water droplets on the FEP-S coating. At the same time, the contact anchorage point was very small, so the ice adhesion strength was low, and the ice drop could be lightly removed from the coating by a slight jerk of a pipette.

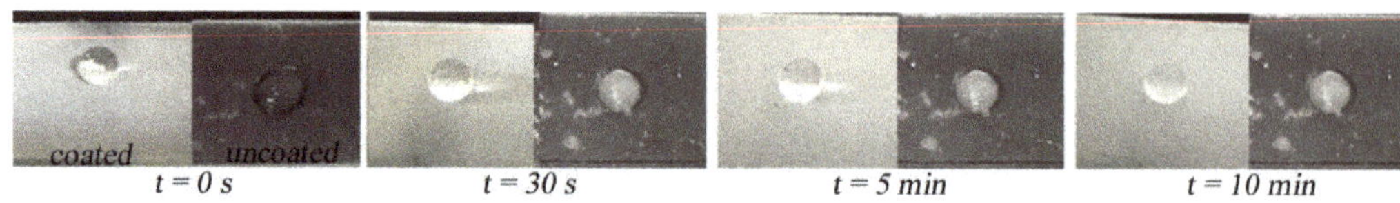

Figure 13. Anti-icing properties of FEP-S coating.

In order to further observe the dynamic icing behavior of the FEP-S coating, we placed the FEP-S coating in the freezer at -18 °C for 24 h, and then dropped 0 °C water drops from the ice water mixture to observe the state of the water droplets. As shown in Video S1, the FEP-S coating placed at -18 °C for 24 h still exhibited good superhydrophobicity. Once the water drops fell on the surface of the coating, it rebound quickly, which means that the water drops bounced off the coating and rolled away before the temperature of the water drops to the freezing point.

The results show that the FEP-S coating had good static and dynamic anti-icing performance in extremely freezing weather. Even in extreme cold conditions, the coating still had superhydrophobic and anti-icing properties, which made the FEP-S coating still have certain self-cleaning and anti-fouling ability.

4. Conclusions

To summarize, we have successfully synthesized a fluorinated epoxy containing an epoxy group and long perfluoroalkyl chains by a thiol-ene click reaction. We have constructed a re-entrant-like structure by adding different proportions of unmodified nano- and micro-silica into the fluorinated epoxy, and obtain the superhydrophobic and oleophobic surface with excellent comprehensive properties. The as-prepared FEP-S surface exhibits high CA and low SA for water, with WCA higher than 158.6° $\pm$ 1.1° and SA lower than 10°, and the CAs for glycerol, ethylene glycol, and diiodomethane reached 152.4° $\pm$ 0.9°, 153.4° $\pm$ 1.3°, and 140.7° $\pm$ 0.9°, respectively.

In addition, the coating exhibited good durability. After sandpaper abrasion for 10 m, FEP-S surface still had good amphiphobicity. The fabricated FEP-S surface exhibits excellent performance in self-cleaning, corrosion resistance, and anti-icing. The coating also showed a remarkable durability towards strong acid and alkali.

Supplementary Materials: The following are available online at https://www.mdpi.com/article/10.3390/coatings11060663/s1, Video S1: Anti-Icing Performance of FEP-S Superhydrophobic Coatings.

Author Contributions: Conceptualization, R.Y.; investigation, X.H.; project administration, R.Y.; software, X.H.; supervision, R.Y.; writing—original draft, X.H. All authors have read and agreed to the published version of the manuscript.

Funding: This research received no external funding.

Institutional Review Board Statement: Not applicable.

Informed Consent Statement: Not applicable.

Data Availability Statement: Data sharing not applicable. No new data were created or analyzed in this study. Data sharing is not applicable to this article.

Conflicts of Interest: The authors declare no conflict of interest.

References

1. Zhang, H.; Ma, Y.; Tan, J.J.; Fan, X.L.; Liu, Y.B.; Gu, J.W.; Zhang, B.L.; Zhang, H.P.; Zhang, Q.Y. Robust, self-healing, superhydrophobic coatings highlighted by a novel branched thiol-ene fluorinated siloxane nanocomposites. *Compos. Sci. Technol.* **2016**, *137*, 78–86. [CrossRef]
2. Chen, J.H.; Liu, Z.H.; Wen, X.F.; Xu, S.P.; Wang, F.; Pi, P.H. Two-Step Approach for Fabrication of Durable Superamphiphobic Fabrics for Self-Cleaning, Antifouling, and On-Demand Oil/Water Separation. *Ind. Eng. Chem. Res.* **2019**, *58*, 5490–5500. [CrossRef]

3.   Wang, H.H.; Lu, H.T.; Zhang, X. Super-robust superamphiphobic surface with anti-icing property. *RSC Adv.* **2019**, *9*, 27702–27709. [CrossRef]

4.   Li, Y.; Hu, T.; Li, B.; Wei, J.; Zhang, J. Totally Waterborne and Highly Durable Superamphiphobic Coatings for Anti-Icing and Anticorrosion. *Adv. Mater. Interfaces* **2019**, *6*. [CrossRef]

5.   Su, F.; Yao, K. Facile fabrication of superhydrophobic surface with excellent mechanical abrasion and corrosion resistance on copper substrate by a novel method. *ACS Appl. Mater. Interfaces* **2014**, *6*, 8762–8770. [CrossRef]

6.   Zhang, B.; Xu, W.; Xia, D.; Huang, Y.; Zhao, X.; Zhang, J. Spray coated superamphiphobic surface with hot water repellency and durable corrosion resistance. *Colloid Surf. A* **2020**, *596*. [CrossRef]

7.   Wang, H.; Zhou, H.; Niu, H.; Zhang, J.; Du, Y.; Lin, T. Dual-Layer Superamphiphobic/Superhydrophobic-Oleophilic Nanofibrous Membranes with Unidirectional Oil-Transport Ability and Strengthened Oil-Water Separation Performance. *Adv. Mater. Interfaces* **2015**, *2*. [CrossRef]

8.   Wang, H.; Zhang, Z.; Wang, Z.; Liang, Y.; Cui, Z.; Zhao, J.; Li, X.; Ren, L. Multistimuli-Responsive Microstructured Superamphiphobic Surfaces with Large-Range, Reversible Switchable Wettability for Oil. *ACS Appl. Mater. Inter.* **2019**, *11*, 28478–28486. [CrossRef]

9.   Tuteja, A.; Choi, W.; Ma, M.; Mabry, J.M.; Mazzella, S.A.; Rutledge, G.C.; McKinley, G.H.; Cohen, R.E. Designing superoleophobic surfaces. *Science* **2007**, *318*, 1618–1622. [CrossRef]

10.   Telecka, A.; Li, T.; Ndoni, S.; Taboryski, R. Nanotextured Si surfaces derived from block-copolymer self-assembly with superhydrophobic, superhydrophilic, or superamphiphobic properties. *RSC Adv.* **2018**, *8*, 4204–4213. [CrossRef]

11.   Tuominen, M.; Teisala, H.; Haapanen, J.; Makela, J.M.; Honkanen, M.; Vippola, M.; Bardage, S.; Walinder, M.E.P.; Swerin, A. Superamphiphobic overhang structured coating on a biobased material. *Appl. Surf. Sci.* **2016**, *389*, 135–143. [CrossRef]

12.   Huang, L.; Yao, Y.; Peng, Z.; Zhang, B.; Chen, S. How to Achieve a Monostable Cassie State on a Micropillar-Arrayed Superhydrophobic Surface. *J. Phys. Chem. B* **2021**, *125*, 883–894. [CrossRef]

13.   Zhu, C.; Gao, Y.; Huang, Y.; Li, H.; Meng, S.; Francisco, J.S.; Zeng, X.C. Controlling states of water droplets on nanostructured surfaces by design. *Nanoscale* **2017**, *9*, 18240–18245. [CrossRef] [PubMed]

14.   Chen, L.; Guo, Z.; Liu, W. Biomimetic Multi-Functional Superamphiphobic FOTS-TiO$_2$ Particles beyond Lotus Leaf. *ACS Appl. Mater. Inter.* **2016**, *8*, 27188–27198. [CrossRef]

15.   Yao, W.; Li, L.; Li, O.L.; Cho, Y.-W.; Jeong, M.-Y.; Cho, Y.-R. Robust, self-cleaning, amphiphobic coating with flower-like nanostructure on micro-patterned polymer substrate. *Chem. Eng. J.* **2018**, *352*, 173–181. [CrossRef]

16.   Li, H.; Yu, S.; Han, X. Fabrication of CuO hierarchical flower-like structures with biomimetic superamphiphobic, self-cleaning and corrosion resistance properties. *Chem. Eng. J.* **2016**, *283*, 1443–1454. [CrossRef]

17.   Velayi, E.; Norouzbeigi, R. Single-step prepared hybrid ZnO/CuO nanopowders for water repellent and corrosion resistant coatings. *Ceram. Int.* **2019**, *45*, 16864–16872. [CrossRef]

18.   Ge, D.; Yang, L.; Zhang, Y.; Rahmawan, Y.; Yang, S. Transparent and Superamphiphobic Surfaces from One-step Spray Coating of Stringed Silica Nanoparticle/Sol Solutions. *Part. Part. Syst. Char.* **2014**, *31*, 763–770. [CrossRef]

19.   Zhi, S.; Wang, G.; Zeng, Z.; Zhu, L.; Liu, Z.; Zhang, D.; Xu, K.; Xue, Q. 3D mossy structures of zinc filaments: A facile strategy for superamphiphobic surface design. *J. Colloid Interf. Sci.* **2018**, *526*, 106–113. [CrossRef]

20.   Hyuneui, L. Beyond a nature-inspired lotus surface: Simple fabrication approach part I. Superhydrophobic and transparent biomimetic glass part II. Superamphiphobic web of nanofibers. In *Advances in Biomimetics*; InTech Open: London, UK, 2011; pp. 145–158.

21.   Deng, X.; Mammen, L.; Butt, H.J.; Vollmer, D. Candle soot as a template for a transparent robust superamphiphobic coating. *Science* **2012**, *335*, 67–70. [CrossRef]

22.   Liu, Z.J.; Wang, H.Y.; Wang, W.Q.; Zhang, X.G.; Yuan, R.X.; Zhu, Y.J. Superhydrophobic poly(vinylidene fluoride) membranes with controllable structure and tunable wettability prepared by one-step electrospinning. *Polymer* **2016**, *82*, 105–113. [CrossRef]

23.   Ganesh, V.A.; Dinachali, S.S.; Nair, A.S.; Seeram, R. Robust Superamphiphobic Film from Electrospun TiO$_2$ Nanostructures. *ACS Appl. Mater. Inter.* **2013**, *5*, 1527–1532. [CrossRef]

24.   Li, T.; Paliy, M.; Wang, X.; Kobe, B.; Lau, W.-M.; Yang, J. Facile One-Step Photolithographic Method for Engineering Hierarchically Nano/Microstructured Transparent Superamphiphobic Surfaces. *ACS Appl. Mater. Inter.* **2015**, *7*, 10988–10992. [CrossRef]

25.   Dong, S.; Zhang, X.; Li, Q.; Liu, C.; Ye, T.; Liu, J.; Xu, H.; Zhang, X.; Liu, J.; Jiang, C.; et al. Springtail-Inspired Superamphiphobic Ordered Nanohoodoo Arrays with Quasi-Doubly Reentrant Structures. *Small* **2020**, *16*. [CrossRef]

26.   Stefelová, J.; Slovák, V.; Siqueira, G.; Olsson, R.T.; Tingaut, P.; Zimmermann, T.; Sehaqui, H. Drying and Pyrolysis of Cellulose Nanofibers from Wood, Bacteria, and Algae for Char Application in Oil Absorption and Dye Adsorption. *ACS Sustain. Chem. Eng.* **2017**, *5*, 2679–2692. [CrossRef]

27.   Yin, K.; Dong, X.; Zhang, F.; Wang, C.; Duan, J.A. Superamphiphobic miniature boat fabricated by laser micromachining. *Appl. Phys. Lett.* **2017**, *110*. [CrossRef]

28.   Wan, Y.L.; Chuan, W.X.; Liu, Z.G.; Yu, H.D.; Li, J. Preparation of superamphiphobic aluminium alloy surface based on laser-EDM method. *Micro Nano Lett.* **2018**, *13*, 281–283. [CrossRef]

29.   Yin, K.; Du, H.F.; Luo, Z.; Dong, X.R.; Duan, J.A. Multifunctional micro/nano-patterned PTFE near-superamphiphobic surfaces achieved by a femtosecond laser. *Surf. Coat. Technol.* **2018**, *345*, 53–60. [CrossRef]

30. Sun, Y.W.; Wang, L.L.; Gao, Y.Z.; Guo, D.M. Preparation of stable superamphiphobic surfaces on Ti-6Al-4V substrates by one-step anodization. *Appl. Surf. Sci.* **2015**, *324*, 825–830. [CrossRef]

31. Barthwal, S.; Kim, Y.S.; Lim, S.-H. Fabrication of amphiphobic surface by using titanium anodization for large-area three-dimensional substrates. *J. Colloid Interf. Sci.* **2013**, *400*, 123–129. [CrossRef]

32. Zhang, J.; Yu, B.; Wei, Q.; Li, B.; Li, L.; Yang, Y. Highly transparent superamphiphobic surfaces by elaborate microstructure regulation. *J. Colloid Interf. Sci.* **2019**, *554*, 250–259. [CrossRef]

33. Ellinas, K.; Pujari, S.P.; Dragatogiannis, D.A.; Charitidis, C.A.; Tserepi, A.; Zuilhof, H.; Gogolides, E. Plasma Micro-Nanotextured, Scratch, Water and Hexadecane Resistant, Superhydrophobic, and Superamphiphobic Polymeric Surfaces with Perfluorinated Monolayers. *ACS Appl. Mater. Inter.* **2014**, *6*, 6510–6524. [CrossRef]

34. Cai, J.; Wang, T.; Hao, W.; Ling, H.; Hang, T.; Chung, Y.-W.; Li, M. Fabrication of superamphiphobic Cu surfaces using hierarchical surface morphology and fluorocarbon attachment facilitated by plasma activation. *Appl. Surf. Sci.* **2019**, *464*, 140–145. [CrossRef]

35. Ellinas, K.; Tserepi, A.; Gogolides, E. Superhydrophobic Fabrics with Mechanical Durability Prepared by a Two-Step Plasma Processing Method. *Coatings* **2018**, *8*, 351. [CrossRef]

36. Wang, H.; Liu, Z.; Wang, E.; Zhang, X.; Yuan, R.; Wu, S.; Zhu, Y. Facile preparation of superamphiphobic epoxy resin/modified poly(vinylidene fluoride)/fluorinated ethylene propylene composite coating with corrosion/wear-resistance. *Appl. Surf. Sci.* **2015**, *357*, 229–235. [CrossRef]

37. Li, X.; Li, H.; Huang, K.; Zou, H.; Yu, D.; Li, Y.; Qiu, B.; Wang, X. Durable superamphiphobic nano-silica/epoxy composite coating via coaxial electrospraying method. *Appl. Surf. Sci.* **2018**, *436*, 283–292. [CrossRef]

38. Yousefi, E.; Ghadimi, M.R.; Amirpoor, S.; Dolad, A. Preparation of new superhydrophobic and highly oleophobic polyurethane coating with enhanced mechanical durability. *Appl. Surf. Sci.* **2018**, *454*, 201–209. [CrossRef]

39. Huang, C.; Wang, F.; Wang, D.; Guo, Z. Wear-resistant and robust superamphiphobic coatings with hierarchical TiO_2/SiO_2 composite particles and inorganic adhesives. *New J. Chem.* **2020**, *44*, 1194–1203. [CrossRef]

40. Xiong, D.; Liu, G.; Duncan, E.J.S. Robust amphiphobic coatings from bi-functional silica particles on flat substrates. *Polymer* **2013**, *54*, 3008–3016. [CrossRef]

41. Su, C. A simple and cost-effective method for fabricating lotus-effect composite coatings. *J. Coat. Technol. Res.* **2012**, *9*, 135–141. [CrossRef]

42. Zhang, F.; Qian, H.; Wang, L.; Wang, Z.; Du, C.; Li, X.; Zhang, D. Superhydrophobic carbon nanotubes/epoxy nanocomposite coating by facile one-step spraying. *Surf. Coat. Technol.* **2018**, *341*, 15–23. [CrossRef]

43. Xiu, Y.H.; Liu, Y.; Balu, B.; Hess, D.W.; Wong, C. Robust Superhydrophobic Surfaces Prepared with Epoxy Resin and Silica Nanoparticles. *IEEE T. Comp. Pack. Man.* **2012**, *2*, 395–401. [CrossRef]

44. Han, X.; Peng, J.; Jiang, S.; Xiong, J.; Song, Y.; Gong, X. Robust Superamphiphobic Coatings Based on Raspberry-like Hollow SnO_2 Composites. *Langmuir* **2020**, *36*, 11044–11053. [CrossRef] [PubMed]

45. Peng, J.; Zhao, X.; Wang, W.; Gong, X. Durable Self-Cleaning Surfaces with Superhydrophobic and Highly Oleophobic Properties. *Langmuir* **2019**, *35*, 8404–8412. [CrossRef]

46. Aslanidou, D.; Karapanagiotis, I.; Panayiotou, C. Tuning the wetting properties of siloxane-nanoparticle coatings to induce superhydrophobicity and superoleophobicity for stone protection. *Mater. Des.* **2016**, *108*, 736–744. [CrossRef]

47. Zhu, K.; Zhang, J.; Zhang, H.; Tan, H.; Zhang, W.; Liu, Y.; Zhang, H.; Zhang, Q. Fabrication of durable superhydrophobic coatings based on a novel branched fluorinated epoxy. *Chem. Eng. J.* **2018**, *351*, 569–578. [CrossRef]

48. Zhang, H.; Tan, J.; Liu, Y.; Hou, C.; Ma, Y.; Gu, J.; Zhang, B.; Zhang, H.; Zhang, Q. Design and fabrication of robust, rapid self-healable, superamphiphobic coatings by a liquid-repellent "glue plus particles" approach. *Mater. Des.* **2017**, *135*, 16–25. [CrossRef]

49. Wang, K.L.; Liu, X.R.; Tan, Y.; Zhang, W.; Zhang, S.F.; Liu, J.Z.; Huang, A.M. Highly fluorinated and hierarchical HNTs/SiO₂ hybrid particles for substrate-independent superamphiphobic coatings. *Chem. Eng. J.* **2018**, *359*, 626–640. [CrossRef]

50. Li, Y.B.; Li, B.C.; Zhao, X.; Tian, N.; Zhang, J.P. Totally Waterborne, Nonfluorinated, Mechanically Robust, and Self-Healing Superhydrophobic Coatings for Actual Anti-Icing. *ACS Appl. Mater. Inter.* **2018**, *10*, 39391–39399. [CrossRef]

Viewpoint

Critical Aspects in Fabricating Multifunctional Super-Nonwettable Coatings Exhibiting Icephobic and Anti-Biofouling Properties

Karekin D. Esmeryan

Acoustoelectronics Laboratory, Georgi Nadjakov Institute of Solid State Physics, Bulgarian Academy of Sciences, 72, Tzarigradsko Chaussee Blvd., 1784 Sofia, Bulgaria; karekin_esmerian@abv.bg; Tel.: +359-2-979-5811

Abstract: The water is a vital compound for all known forms of life, but it can also cause detrimental consequences to our daily routine if by natural means becomes pathogenic bacterial carrier or transforms into ice. Imaginative by necessity, the surrounding environment has stimulated the mankind to emulate natural-design solutions and invent the so-called super-nonwettable coatings. Undisputedly, these coatings have revolutionized the modern industry by providing "a vehicle" for potential eco-friendly water purification, passive icing protection, suppression of the solid surface-associated spreading of bacterial infections and enhanced cryopreservation of living matter. Regrettably, the wide domestic use of liquid impermeable coatings (surfaces) is yet limited, since the current market trends impose the possession of fabrication scalability and multifunctionality, which is not covered by most of the available non-wettable products. This viewpoint article intends to outline the most significant scientific achievements within the past five years related to the release of anti-wetting coatings with multiple applications. Design and performance efficiencies in light of the physical chemistry of the surface are demonstrated, emphasizing on the likelihood of integrating icephobicity and anti-biofouling capacity within a single interfacial nanostructure.

Keywords: multifunctional; carbon soot coatings; super-nonwettable

Citation: Esmeryan, K.D. Critical Aspects in Fabricating Multifunctional Super-Nonwettable Coatings Exhibiting Icephobic and Anti-Biofouling Properties. *Coatings* **2021**, *11*, 339. https://doi.org/10.3390/coatings11030339

Academic Editor: N. P. Prorokova

Received: 1 March 2021
Accepted: 13 March 2021
Published: 16 March 2021

Publisher's Note: MDPI stays neutral with regard to jurisdictional claims in published maps and institutional affiliations.

1. Introduction

For decades, the perfect symbiosis among the physical, chemical and biological processes in our ecosystem has served as a driving force to the human race for life-changing scientific inventions and technological progress [1]. An instance of classy biomimicry is the currently intensive fabrication of liquid-repellent coatings, directly provoked by the eternally dry and clean leaves of the sacred Lotus [2].

The aim of endowing a surface-of-interest with exceptional resistance to wetting is dictated by the necessity of minimizing the solid-liquid contact area and promoting the creation of self-cleaning and drag-reducing surfaces, passive anti-icing systems, anti-microbial and anti-corrosive coatings, enhanced pervaporative separation membranes, water purification devices, biological sensors for human semen quality analysis or new technologies for cryopreservation of living matter [3–5]. For instance, it is crucially important to keep different industrial facilities such as high-voltage power lines or telecommunication towers free of ice/snow, since the ice/snow accretion may block their operability and thus the people's access to electricity, telecommunications and internet [6,7]. In parallel, up to 26% of all hospital-associated infections occur via biofouling of the medical equipment, which can be alleviated using coatings with custom surface topographies [8,9]. However, the increasing demand of economic profitability and scalability of a given commercial anti-wetting coating implies that the latter needs to be flexible and combine several functionalities. Therefore, the global scientific community has steadily shifted its recent efforts towards designing multifunctional super-nonwettable coatings [10–12].

The inherent meaning of "multifunctional" implies simultaneous possession of more than two relevant properties with different physicochemical origin (e.g., icephobicity, anti-bioadhesiveness, self-healing, etc.). Regrettably, in many cases this expression is somehow inappropriately used to take into account the self-cleaning and anti-corrosion capabilities of the surface [13–15]. Its non-wetting state is interrelated to the above-mentioned features, since the weak interfacial attraction forces lead to liquid droplet contact angle exceeding 150°, contact angle hysteresis below 5° and rolling angle less than 10°, converting the substrate to quasifrictionless and forming a slippery interface [16–20]. Despite being dually roughened, hierarchical or composed of randomly distributed micro- nanoparticles, every super-nonwettable surface is expected to exhibit self-cleaning and to a certain extent anti-corrosion performance [14]. In contrast, and although the suppression of the ice nucleation and biofilm spreading could be governed by the degree of water repellency [21,22], the contrivance of anti-icing and anti-microbial coatings is not a straightforward procedure and requires careful consideration of some additional parameters [4,23].

The idea of this viewpoint article is to serve as a useful information platform for those scholars working in the field of wetting phenomena, whose research is dedicated to the fabrication of multifunctional non-wettable coatings. Thus, the discussion is focused on the main principles and difficulties related to the design of universal and versatile liquid-repellent interfaces, considering the findings described in specially-selected scientific publications reporting multifarious surface properties. A few hypotheses concerning the future incorporation of icephobic and anti-bioadhesive characteristics within a single functional coating are presented too in the context of well-established physicochemical mechanisms.

2. Importance of the Physicochemical Profile

A significant factor contributing to the development of diversified non-wettable thin films is the choice of appropriate chemicals and roughness modification protocols, since the appearance of hydrophobic micro-nanoprotrusions reduces the interfacial contact, leading to weaker solid-liquid attraction per unit area (compared to a flat surface with the same non-polar chemistry) and amplification of the hydrophobic effect [16]. While combining non-polar chemistry and micro-nanoscale surface asperities is suitable for anti-wetting applications [16], the assignment of fluorine-containing compounds may lead to superamphiphobicity (repelling both water and oil) [24], impeding the oil filtration efficiency of the interface. Therefore, many researchers prefer to employ fluorine-free strategies for surface patterning, because they ensure non-wettability to a variety of liquids such as water, glycerol, ethylene glycol, blood, tea, juice, milk, etc., but the surface is still wettable by oils (decisive when developing oil-water separation membranes) and the fabrication is environmentally benign [10–12,25–28]. Commonly used chemicals are carboxymethyl cellulose (CMC), hydroxyapatite nanowires (HAP), ZnO nanoparticles [10], fly ash [12,26], dymethyloctadecyl ammonium chloride [7], TiO_2-PDMS [12,28], tetrabutyl-titanate-dymethylformamide-acetic acid-n Octyltrietoxysilane [25] or tetraethoxysilane hexamethyldisilazane [27], whose different combinations may endow superior mechanical durability of the coated surface and "a bouquet" of functionalities.

Embedding fly ash or activated carbon in the matrix of TiO_2-PDMS is demonstrated to render self-healing, oil-water separation, UV protective, photocatalytic and even flame-retardant properties of the coatings [12,28], shown concisely in Figures 1 and 2, which is a good alternative to the more sophisticated $CuFe_2O_4$-TiO_2 nanocomposite supported reduced graphene oxide photocatalysts [29]. It is argued that the self-healing ability is due to self-migration of PDMS molecules towards the surface by heating, leading to recovered superhydrophobicity [28]. In comparison, the flame-retardancy is attributed to the incombustibility of the fly ash and its covalent bonding to the TiO_2-PDMS [12].

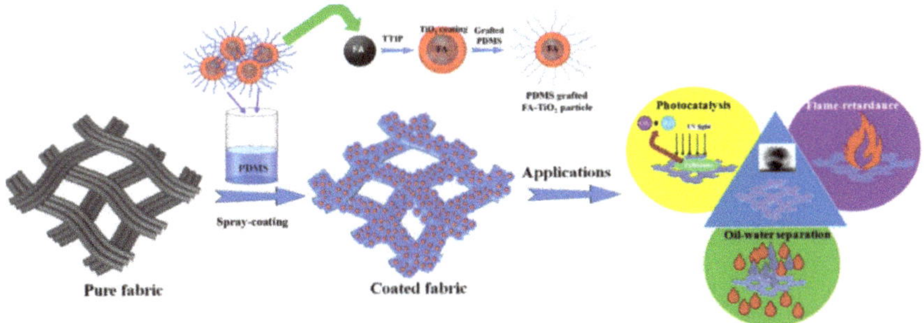

Figure 1. Multifunctional properties of a fly ash-TiO$_2$-PDMS spray coated superhydrophobic fiber. Reproduced with permission from [12]. Copyright 2020 Elsevier.

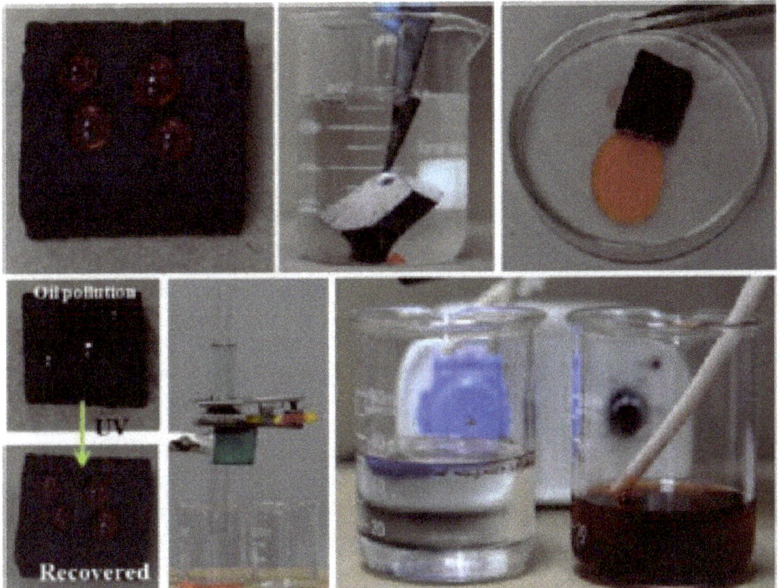

Figure 2. Illustration of the applicability of an activated carbon-TiO$_2$-PDMS coated superhydrophobic sponge. Reproduced with permission from [28]. Copyright 2020 ACS Publications.

On the other hand, the resistivity to burning and the self-healing capacity can be gained by adding inherently fire-resistant hydroxyapatite and thermally stable ZnO, which also yields a bactericidal effect of the solid surface [10], revealed in Figure 3.

Figure 3. Bactericidal effect of modifier-free superhydrophobic and flame-resistant paper. Reproduced with permission from [10]. Copyright 2018 Elsevier.

The surface porosity, morphology and topography are another three parameters of practical significance in terms of the anti-microbial, oil- and sound absorption effectiveness of the interface [27,30]. An innovative technique of adjusting them is via compression and strict monitoring of the curing temperature, since in such a way the object's thickness and particles' size are accurately controlled [27]. Namely, Figure 4 depicts how a superhydrophobic latex sponge is fabricated by immersing it into a preliminary prepared colloidal silica sol composed of tetraethoxysilane, hexamethyldisilazane, ethanol and de-ionized water, further homogeneously mixed with epoxy resin and polyamide by stirring. The sponge is then dried in air for 24 h and during this procedure, the samples could be free-standing or strongly bonded to the substrate depending on whether the specimens are flipped or no [27]. Upon compression, a sheet-like superhydrophobic material with desired thickness might be obtained. Meanwhile, if epoxy spherical structures with dimensions exceeding 5 μm are required, the curing temperature should be 15–25 °C, while at 60 °C, the size decreases below 2 μm. The authors explain these observations with the epoxy resin's solubility, which enhances at high temperatures, inducing shrinking of the gel [27]. Moreover, as described elsewhere [25], the hot pressing of cotton fabrics at 200 °C leads to an UV protection factor of 45 (10 times higher than pristine cotton fabrics) owing to the undergone crystal phase transition. Furthermore, tuning the quantity of the chemical components in any coating would reflect on the corresponding wettability, filtration characteristics and UV protection, based on the observations in ref. [10]. The excessive concentration of ZnO, for example, increases the number of hydrophilic surface sites and the superhydrophobicity and UV resistivity are degraded due to the formation of oxygen vacancies resulting from electron-hole pairs generation and reaction of the holes with the oxygen sites [10], which may trigger the absorption of UV-light-forming excitons [31].

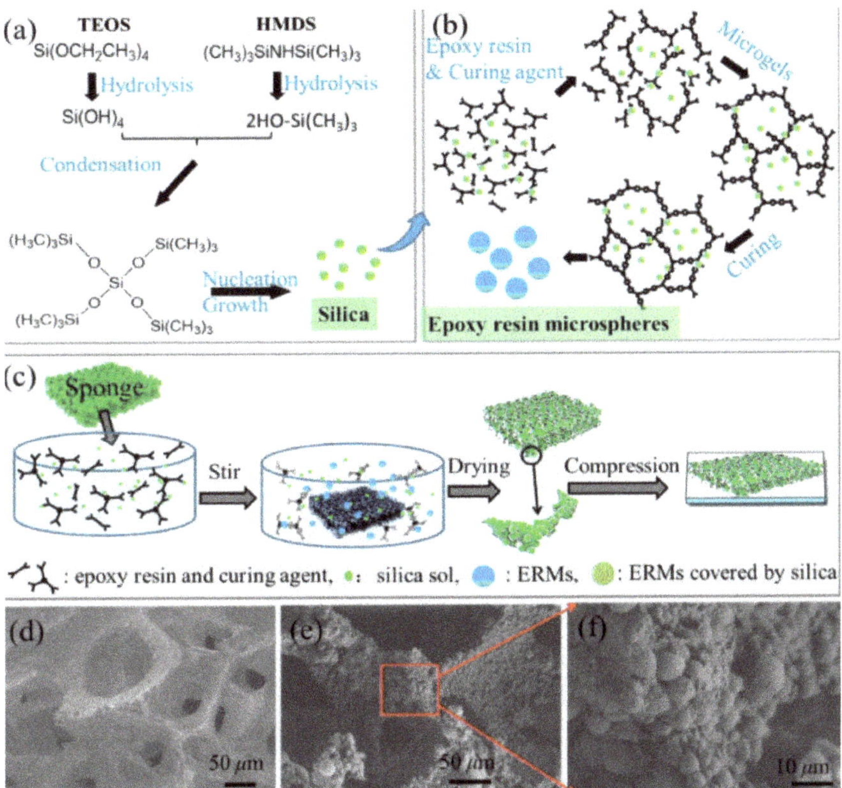

Figure 4. Formation mechanism, synthesis and morphology of water-repellent latex sponges. (**a**) Speculated formation pathway of silica. (**b**) Speculated formation pathway of ERMs. (**c**) Schematic illustration of the one-pot synthesis procedure of ERMs@Latex-T. SEM images of latex porous structures before (**d**) and after (**e**) ERMs adhering. (**f**) The high magnification image of (**e**). The reaction temperature of the sample was 25 °C. Reproduced with permission from [27]. Copyright 2019 Elsevier.

Altogether, to engineer liquid-repellent materials manifesting contemporaneously more than two functional properties, one needs to ensure that the synthesis method will provide a coating with non-polar chemical composition and micro-nanoscale surface roughness, thus, amplifying the effect of hydrophobicity and minimizing the solid-liquid interactions [16]. In the meantime, the interface has to maintain photocatalytic activity upon UV irradiation (e.g., for decomposing the absorbed organic pollutants) or self-replenishment of the lost/disintegrated hydrophobic molecules for sustainable long-term non-wettability. The current state-of-the-art in the field proposes TiO_2 and PDMS as suitable compounds fulfilling the above requirements [12,28], however, it will be wise to look for existing alternatives including silver gallium sulfide ($AgGaS_2$) and zinc sulfide (ZnS) due to the proven bactericidal behavior of Ag and Zn. Besides, it is pertinent to attentively consider the scalability of the experimental approach, since the industrial physicochemical processing must be feasible at large scales, irrelevantly of the complexity and geometry of the solid substrate. Nowadays, the combustion flame synthesis, dip coating and spray deposition are three prominent techniques with undoubtful fabrication scalability [11,12,25–28,32].

3. Icephobicity and Anti-Bioadhesiveness, Are They Unifiable?

At present, and despite the exponential progress of nanotechnologies, the scientific literature seems to suffers from the profound lack of even speculative assumptions on how to design super-nonwettable surfaces, possibly unifying anti-icing and anti-microbial capabilities. On one hand, there is substantial disagreement within the scientific community about the similarities between water and ice repellency, mainly because of the complexity of the heterogeneous ice nucleation, involving the impact of the surface morphology/topography, porosity, roughness, air cushion (plastron) distribution, degree of non-wettability and film thickness [4,21,33]. At the same time, the development of interfaces non-stickable to living biomass (bacteria, fungi, algae, plants, biomolecules, etc.) is very challenging. It depends on the type of microorganisms, their size, shape and cell walls' charge, as well as on the optimal nanopillar radius, spacing and length determining the cell's binding affinity and eventual mechanical rupture [30,34,35]. Interestingly, the super-nonwettable coatings may mitigate the icing by reducing the heat transfer rate and weakening the ice adhesion strength (if the phase transition occurs) as a consequence of the negligible solid fraction in contact with the freezing water and the availability of trapped air gaps [4]. Instead, the biomass attachments are suppressed by increasing the substrate-to-foulant separation distance and the kinetic threshold for bioadhesion [36] or rupturing the cells through the tensile stress exerted by the nanopillars [35]. In other words, these are implicitly distinct mechanisms, whose intersection is unclear so far and the author will try to briefly hypothesize on the integration of icephobicity and anti-bioadhesiveness in an individual coating.

One of the rare cases where the aforementioned anti-icing and anti-microbial pathways are noticed independently of one another refers to the carbon soot, as illustrated in Figures 5 and 6.

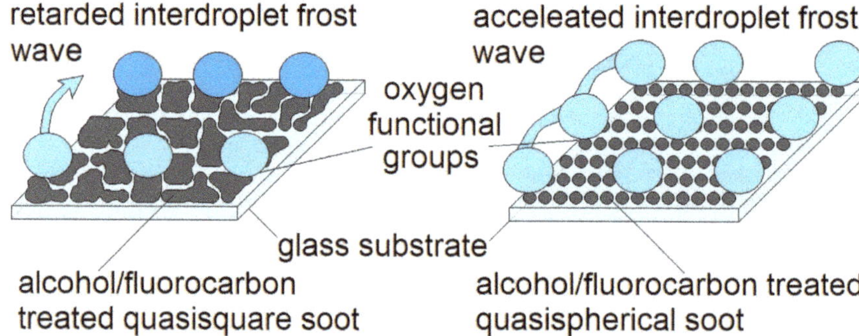

Figure 5. Delaying the interdroplet ice bridging using super-nonwettable carbon soot coatings. The opaque blue and sky blue droplets represent, the frozen and unfrozen subcooled condensates, respectively. The quasisquares (**left side**) and spheres (**right side**) denote the surface morphology of the soot. The lines below the frozen droplets aim to depict the oxygen functional groups serving as ice nucleation (active) sites. The abundance of active sites on the quasispherical soot (**right side**) accelerates the freezing events and the velocity of the interdroplet freezing [37].

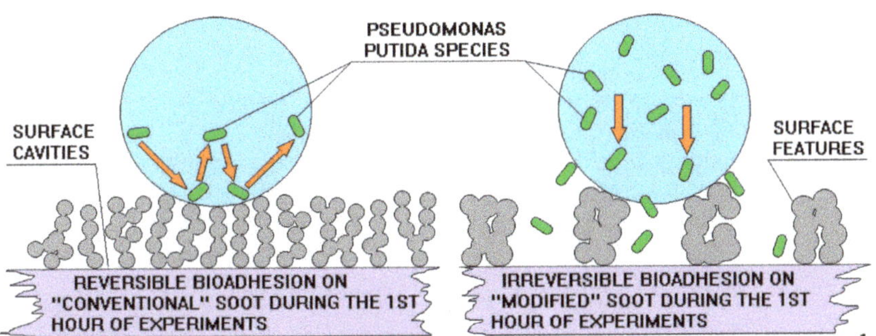

Figure 6. Reversible (**left part**) and irreversible (**right part**) bioadhesion on super-nonwettable carbon soot coatings. Reproduced with permission from [24]. Copyright 2020 Elsevier.

Some newest studies divulge that the incipiency of ice nuclei on the soot is regulated by the amount and spatial arrangement of its oxygen active sites serving as nucleation centers that regulate the hydrogen bonding with the surrounding water vapor and the subsequent interdroplet frost wave [37]. It is found that the degree of soot oxidation and the space between the individual active sites can be successfully manipulated via alcohol modifications causing retardation or acceleration of the freezing front velocity regardless of the nanoparticles' size and shape (i.e., the surface morphology) [37]. However, if the surface roughness curvature (the nanocavities associated with the roughening) is closer or larger than the critical nucleus' radius, the thermodynamic barrier for heterogeneous nucleation might still be decreased despite the material's weak oxidation [21]. Conversely, the bacterial accumulation on sooted substrates is highly sensitive to the soot particle/aggregate morphology and if the dimensions of the surface protrusions and the distance among them are commensurable with the length scale of the adhering living matter, the solid-cell contact area increases, morphologically-triggered partial wetting occurs and the cells binding becomes irreversible [30,34]. Obviously, at a first glance, the icephobicity and the anti-bioadhesiveness are functional features with no identifiable abutment, so it is quite provocative and complicated to invent an interfacial coating with synergistic ice repellent and anti-bioadhesion performance.

Providing a more general view, because this is the primary objective herein, a universal surface (coating) accommodating the above properties may have to consist of solid particles with as small as possible size (nanometer scale) in order to increase the Gibbs thermodynamic free energy barrier and limit the locally-wetted solid fractions. Hence, the liquid-solid phase transitions would be delayed up to a few hours [38] and if immersed in biocontaminated solutions, the surface could stay free of attached biomass [30,34]. Other mandatory requirement might be to set an aggregate-to-aggregate (pillar-to-pillar) distance lower than the size of the biospecies, but large enough to ensure the appearance of steady air gaps (i.e., plastron), enhancing the thermal resistance of the material (slowing the heat transfer rate) and guaranteeing sufficient tensile stress experienced by the colonizing microorganisms, as a result of the sagging among the protuberances, and their eventual physical damage [35]. Finally, avoiding the presence of oxygen functional groups on the surface is highly beneficial for preventing the hydrogen bonds formation and canceling the chemistry-mediated freezing events [37].

Quantitatively, suppose the target is to inhibit the nascency of ice embryos at negative temperatures and simultaneously regulating the spontaneous adherence of Gram-positive (e.g., *Escherichia coli*) or Gram-Negative (e.g., *Pseudomonas putida*) bacterial strains. In that case, and taking into account the size of the strains [34], it would hypothetically be wise to employ a laboratory patterning technique allowing the production of pillars with length and diameter less than 800 nm and 500 nm, accordingly, and a pillar pitch of approximately 900–1500 nm. One of the best ways of accomplishing such a task is via photolithography, *"where a layer is illuminated through a patterned mask"*, facilitating the preparation of well-defined surfaces in multiple identical copies [16]. Of course, this is a costly method, so the templating or chemical etching are relatively more cost-effective options [16], additionally to the adjustment of the physicochemical features by immersion in short alkyl chain alcohols and fluorocarbon emulsions, as recently reported for carbon soot and wax-based coatings [37,39]. Finally, and since the soot is a graphite-like material [32,40], the above hypothetical considerations might be applicable also to graphene and related materials, so future research efforts in such a direction seem reasonable.

4. Conclusions and Outlook

The launch of reliable and multifunctional non-wettable materials is shown to go beyond the cliché of applying low surface energy chemical treatment to highly roughened solid substrates. Finding a subtle balance among the particles' morphology, surface topography, porosity and ratio of the chemical reagents during the synthesis procedure is the key for supporting self-replenishing, chemically non-polar interfaces, successfully absorbing and decomposing hazardous organic oil wastes, while exhibiting UV shielding, fire resistivity and anti-microbial behavior. A synergistic anti-icing and anti-bioadhesive performance is speculated to be achievable if special attention is devoted to the opportunities of optimizing the geometrical distribution of the nanopillars making up the surface, because of the necessity of having sustainable air cushion for optimal heat transfer reduction and pillar-induced tensile stress that breaks the cell membrane and kills the attached biospecies. Enriching the knowledge in surface sciences and advancing the development of mechanically robust and broadly applicable liquid impervious surfaces, in author's opinion, goes hand-in-hand with elucidating the mixed impact of the collective physicochemical components; an aspect scantily addressed in the scientific literature up to now.

Funding: This research received no external funding.

Institutional Review Board Statement: Not applicable.

Informed Consent Statement: Not applicable.

Data Availability Statement: The data presented in this study are available on request from the corresponding author.

Conflicts of Interest: The author declares no conflict of interest.

References

1. Rinaldi, A. Naturally better. Science and technology are looking to nature's successful design for inspiration. *EMBO Rep.* **2007**, *8*, 995–999. [CrossRef]
2. Lathe, S.S.; Terashima, C.; Nakata, K.; Fujishima, A. Superhydrophobic surfaces developed by mimicking hierarchical surface morphology of Lotus leaf. *Molecules* **2014**, *19*, 4256–4283. [CrossRef]
3. Valipour, M.N.; Birjandi, F.C.; Sargolzaei, J. Super-non-wettable surfaces: A review. *Colloids Surf. A Physicochem. Eng. Asp.* **2014**, *448*, 93–106. [CrossRef]
4. Esmeryan, K.D. From extremely water-repellent coatings to passive icing protection—Principles, limitations and innovative application aspects. *Coatings* **2020**, *10*, 66. [CrossRef]
5. Esmeryan, K.D.; Lazarov, Y.; Stamenov, G.S.; Chaushev, T.A. When condensed matter physics meets biology: Does superhydrophobicity benefiting the cryopreservation of human spermatozoa? *Cryobiology* **2020**, *92*, 263–266. [CrossRef] [PubMed]
6. He, Q.; Lv, X.-F.; Zhao, X.-T. Overhead transmission lines deicing under different incentive displacement. *J. Appl. Mater.* **2014**, *2014*, 872198. [CrossRef]
7. Mulherin, N.D. Atmospheric icing and communication tower failure in the United States. *Cold Reg. Sci. Technol.* **1998**, *27*, 91–104. [CrossRef]
8. Magill, S.S.; Edwards, J.R.; Bamberg, W.; Beldavs, Z.G.; Dumyati, G.; Kainer, M.A.; Lynfield, R.; Maloney, M.; McAllister-Hollod, L.; Nadle, J.; et al. Multistate point-prevalence survey of health care-associated infections. *N. Engl. J. Med.* **2014**, *370*, 1198–1208. [CrossRef]
9. Ge, X.; Ren, C.; Ding, Y.; Chen, G.; Lu, X.; Wang, K.; Ren, F.; Yang, M.; Wang, Z.; Li, J.; et al. Micro/nano-structured TiO_2 surface with dual-functional antibacterial effects for biomedical applications. *Bioact. Mater.* **2019**, *4*, 346–357. [CrossRef]
10. Wen, G.; Gao, X.; Tian, P.; Zhong, L.; Wang, Z.; Guo, Z. Modifier-free fabrication of durable and multifunctional superhydrophobic paper with thermostability and anti-microbial property. *Chem. Eng. J.* **2018**, *346*, 94–103. [CrossRef]

11. Wang, F.; Pi, J.; Song, F.; Feng, R.; Xu, C.; Wang, X.L.; Wang, Y.Z. A superhydrophobic coating to create multi-functional materials with mechanical/chemical/physical robustness. *Chem. Eng. J.* **2020**, *381*, 122539. [CrossRef]
12. Wang, Y.; Peng, S.; Shi, X.; Lan, Y.; Zeng, G.; Zhang, K.; Li, X. A fluorine-free method for fabricating multifunctional durable superhydrophobic fabrics. *Appl. Surf. Sci.* **2020**, *505*, 144621. [CrossRef]
13. Vorobyev, A.Y.; Guo, C. Multifunctional surfaces produced by femtosecond laser pulses. *J. Appl. Phys.* **2015**, *117*, 033103. [CrossRef]
14. Cho, E.-C.; Chang-Jian, C.-W.; Chen, H.-C.; Chuang, K.-S.; Zheng, J.-H.; Hsiao, Y.-S.; Lee, K.-C.; Huang, J.-H. Robust multifunctional superhydrophobic coatings with enhanced water/oil separation, self-cleaning, anti-corrosion and anti-biological adhesion. *Chem. Eng. J.* **2017**, *314*, 347–357. [CrossRef]
15. Wu, L.; Wang, L.; Guo, Z.; Luo, J.; Xue, H.; Gao, J. Durable and multifunctional superhydrophobic coatings with excellent joule heating and electromagnetic interference shielding performance for flexible sensing electronics. *ACS Appl. Mater. Interfaces* **2019**, *11*, 34338–34347. [CrossRef] [PubMed]
16. Shirtcliffe, N.J.; McHale, G.; Atherton, S.; Newton, M.I. An introduction to superhydrophobicity. *Adv. Colloid Interface Sci.* **2010**, *161*, 124–138. [CrossRef]
17. Drelich, J.W.; Boinovich, L.; Chibowski, E.; Volpe, C.D.; Holysz, L.; Marmur, A.; Siboni, S. Contact angles: History of over 200 years of open questions. *Surf. Innov.* **2020**, *8*, 3–27. [CrossRef]
18. Bormashenko, E. Physics of solid-liquid interfaces: From the Young equation to the superhydrophobicity (Review article). *Low Temp. Phys.* **2016**, *42*, 622. [CrossRef]
19. Schmitt, M.; Heib, F. A more appropriate procedure to measure and analyse contact angles/drop shape behaviors. In *Advances in Contact Angle, Wettability and Adhesion III*; Mittal, K.L., Ed.; Scrivener Publishing LLC: Beverly, MA, USA, 2018.
20. Amirfazli, A.; Neumann, A.W. Status of the three-phase line tension: A review. *Adv. Colloid Interface Sci.* **2004**, *110*, 121–141. [CrossRef] [PubMed]
21. Schutzius, T.M.; Jung, S.; Maitra, T.; Eberle, P.; Antonini, C.; Stamatopoulos, C.; Poulikakos, D. Physics of icing and rational design of surfaces with extraordinary icephobicity. *Langmuir* **2015**, *31*, 4807–4821. [CrossRef]
22. Li, B.; Ouyang, Y.; Haider, Z.; Zhu, Y.; Qiu, R.; Hu, S.; Niu, H.; Zhang, Y.; Chen, M. One-step electrochemical deposition leading to superhydrophobic matrix for inhibiting abiotic and microbiologically influenced corrosion of Cu in seawater environment. *Colloids Surf. A Physicochem. Eng. Asp.* **2021**, *616*, 126337. [CrossRef]
23. Vladkova, T.G.; Staneva, A.D.; Gospodinova, D.N. Surface engineered biomaterials and ureteral stents inhibiting biofilm formation and encrustation. *Surf. Coat. Technol.* **2020**, *404*, 126424. [CrossRef]
24. Zhang, B.; Zeng, Y.; Wang, J.; Sun, Y.; Zhang, J.; Li, Y. Superamphiphobic aluminum alloy with low sliding angles and acid-alkali liquids repellency. *Mater. Des.* **2020**, *188*, 108479. [CrossRef]
25. Gao, S.; Huang, J.; Li, S.; Liu, H.; Li, F.; Li, Y.; Chen, G.; Lai, Y. Facile construction of robust fluorine-free superhydrophobic TiO$_2$@fabrics with excellent anti-fouling, water-oil separation and UV-protective properties. *Mater. Des.* **2017**, *128*, 1–8. [CrossRef]
26. Pang, B.; Qian, J.; Zhang, Y.; Jia, Y.; Ni, H.; Pang, S.D.; Liu, G.; Qian, R.; She, W.; Yang, L.; et al. 5S multifunctional intelligent coating with superdurable, superhydrophobic, self-monitoring, self-heating and self-healing properties for existing construction application. *ACS Appl. Mater. Interfaces* **2019**, *11*, 29242–29254. [CrossRef] [PubMed]
27. Huang, Z.-S.; Quan, Y.-Y.; Mao, J.-J.; Wang, Y.-L.; Lai, Y.; Zheng, J.; Chen, Z.; Wei, K.; Li, H. Multifunctional superhydrophobic composite materials with remarkable mechanochemical robustness, stain repellency, oil-water separation and sound-absorption properties. *Chem. Eng. J.* **2019**, *358*, 1610–1619. [CrossRef]
28. Shi, X.; Lan, Y.; Peng, S.; Wang, Y.; Ma, J. Green fabrication of a multifunctional sponge as an absorbent for highly-efficient and ultrafast oil-water separation. *ACS Omega* **2020**, *5*, 14232–14241. [CrossRef]
29. Hafeez, H.Y.; Lakhera, S.K.; Karthik, P.; Anpo, M.; Neppolian, B. Facile construction of ternary CuFe$_2$O$_4$-TiO$_2$ nanocomposite supported reduced graphene oxide (rGO) photocatalysts for the efficient hydrogen production. *Appl. Surf. Sci.* **2018**, *449*, 772–779. [CrossRef]
30. Bruzaud, J.; Tarrade, J.; Celia, E.; Darmanin, T.; de Givenchy, E.T.; Guittard, F.; Herry, J.M.; Guilbaud, M.; Bellon-Fontaine, M.N. The design of superhydrophobic stainless steel surfaces by controlling nanostructures: A key parameter to reduce the implantation of pathogenic bacteria. *Mater. Sci. Eng. C* **2017**, *73*, 40–47. [CrossRef]
31. Schmitt, M. Synthesis and testing of ZnO nanoparticles for photo-initiation: Experimental observation of two different non-migration initiators for bulk polymerization. *Nanoscale* **2015**, *7*, 9532–9544. [CrossRef]
32. Esmeryan, K.D.; Castano, C.E.; Bressler, A.H.; Abolghasemibizaki, M.; Mohammadi, R. Rapid synthesis of inherently robust and stable superhydrophobic carbon soot coatings. *Appl. Surf. Sci.* **2016**, *369*, 341–347. [CrossRef]
33. Sojoudi, H.; Wang, M.; Boscher, N.D.; McKinley, G.H.; Gleason, K.K. Durable and scalable icephobic surfaces: Similarities and distinctions from superhydrophobic surfaces. *Soft Matter* **2016**, *12*, 1938–1963. [CrossRef] [PubMed]
34. Esmeryan, K.D.; Avramova, I.A.; Castano, C.E.; Ivanova, I.A.; Mohammadi, R.; Radeva, E.I.; Stoyanova, D.S.; Vladkova, T.G. Early stage anti-bioadhesion behavior of superhydrophobic soot based coatings towards *Pseudomonas putida*. *Mater. Des.* **2018**, *160*, 395–404. [CrossRef]
35. Watson, G.S.; Green, D.W.; Watson, J.A.; Zhou, Z.; Li, X.; Cheung, G.S.P.; Gellender, M. A simple model for binding and rupture of bacterial cells on nanopillar surfaces. *Adv. Mater. Interfaces* **2019**, *6*, 1801646. [CrossRef]

36. Yoon, S.H.; Rungraeng, N.; Song, W.; Jun, S. Superhydrophobic and superhydrophilic nanocomposite coatings for preventing Escherichia coli K-12 adhesion on food contact surface. *J. Food Eng.* **2014**, *131*, 135–141. [CrossRef]
37. Esmeryan, K.D.; Gyoshev, S.D.; Castano, C.E.; Mohammadi, R. Anti-frosting and defrosting performance of chemically modified super-nonwettable carbon soot coatings. *J. Phys. D Appl. Phys.* **2021**, *54*, 015303. [CrossRef]
38. Boinovich, L.; Emelyanenko, A.M.; Korolev, V.V.; Pashinin, A.S. Effect of wettability on sessile drop freezing: When superhydrophobicity stimulates an extreme freezing delay. *Langmuir* **2014**, *30*, 1659–1668. [CrossRef]
39. Esmaeili, A.E.; Mir, N.; Mohammadi, R. A facile, fast and low-cost method for fabrication of micro/nano-textured superhydrophobic surfaces. *J. Colloid Interface Sci.* **2020**, *573*, 317–372. [CrossRef]
40. Sadezky, A.; Muckenhuber, H.; Grothe, H.; Niessner, R.; Pöschl, U. Raman microspectroscopy of soot and related carbonaceous materials: Spectral analysis and structural information. *Carbon* **2005**, *43*, 1731–1742. [CrossRef]

Article

Development of Multifunctional Coating of Textile Materials Using Silver Microencapsulated Compositions

Luidmila Petrova, Olga Kozlova *, Elena Vladimirtseva *, Svetlana Smirnova *, Anna Lipina and Olga Odintsova *

Department of Chemical Technology of Fibrous Materials, Ivanovo State University of Chemistry and Technology, Sheremetevsky Ave., 7, 153000 Ivanovo, Russia; petrova_ls@isuct.ru (L.P.); prohorova.a94@yandex.ru (A.L.)
* Correspondence: kozlova_ov@isuct.ru (O.K.); vladimirtseva_el@isuct.ru (E.V.); smirnova_sv@isuct.ru (S.S.); odintsova_oi@isuct.ru (O.O.)

Abstract: The efficiency of the method for the synthesis of silver nanoparticles using a system containing oxalic dialdehyde as a reducing agent, and polyguanidine as a stabilizer is shown. An analysis of the data of photon correlation spectroscopy characterizing the sizes of the formed particles in the Ag-polyelectrolyte system is presented. It has been established that the synthesized silver nanoparticles have a stable biocidal effect. The system of biodegradable polyelectrolytes chitosan-xanthan gum for the synthesis of the capsule shell including silver nanoparticles is selected. This will allow the formation of stable polyelectrolyte capsule shells containing oyster mushroom mycelium extract. A protocol for the synthesis of microcapsules by the method of sequential adsorption of chitosan polyelectrolytes and xanthan gum on calcium carbonate templates was developed. Silver nanoparticles are included in the capsule shell, and a biologically active drug (oyster mushroom mycelium extract) is included in the core. The technological mode of complex capsules immobilization on a textile material by the layer-by-layer method is described. The immobilization of multilayer microcapsules on a fibrous substrate is provided by a system of polyelectrolytes: positively charged chitosan and negatively charged xanthan gum. The developed multifunctional coatings make it possible to impart multifunctional properties to textile materials: antibacterial, antimycotic, high hygroscopic properties.

Keywords: cellulose textile material; microencapsulation; antibacterial; antimycotic; wound healing properties; silver; polyelectrolyte microcapsules

check for **updates**

Citation: Petrova, L.; Kozlova, O.; Vladimirtseva, E.; Smirnova, S.; Lipina, A.; Odintsova, O. Development of Multifunctional Coating of Textile Materials Using Silver Microencapsulated Compositions. *Coatings* **2021**, *11*, 159. https://doi.org/10.3390/coatings11020159

Academic Editor: Csaba Balázsi
Received: 29 December 2020
Accepted: 25 January 2021
Published: 29 January 2021

Publisher's Note: MDPI stays neutral with regard to jurisdictional claims in published maps and institutional affiliations.

1. Introduction

Nanosilver is a universal biocide, the particles of which suppress the pathological effect of a wide range of microorganisms, including viruses. Modern methods for the formation of silver-containing composites in natural [1] and synthetic [2] matrices by direct impregnation of nanoparticles into the system, which exhibit high bacteriostatic (inhibition of bacterial growth) or bactericidal (destruction of inoculated bacteria) activity, are shown.

The new approach proposed by the authors is based on the inclusion of silver nanoparticles into the shell of a microcapsule containing a medicinal or biologically active compound.

Microencapsulation of drugs provides prolonged action and safety for humans due to the possibility of using minimum concentrations of the active substance and its controlled release [3]. The introduction of a biologically active or medicinal compound into the capsule core will make it possible to create medical textile materials with multifunctional properties: antibacterial, antimycotic, and wound healing.

Microencapsulation is widely used in various industries. In agriculture and in everyday life, microencapsulated insecticides are used, microcapsules with vitamins, essential and fatty oils are included in various cosmetics (creams, gels, serums), microencapsulated probiotics are used in feed and feed additives in veterinary medicine. The introduction of a new technology for microencapsulation of textile materials will make it possible to obtain a completely unique product with innovative quality and functional indicators.

Microencapsulation is the process of encapsulating a functional substance in a shell that protects it from evaporation, pollution, and the influence of other environmental influences and allows the substance to be released in a prolonged manner [4,5].

Depending on the thickness and material of the shell, the core contents can be released through changes in temperature or pH, biodegradation, etc., [6,7].

The existing microencapsulation methods are conventionally divided into chemical, physical [8], and physicochemical [9,10]. When choosing the most suitable method for each specific case, one proceeds from the specified properties of the final product, the cost of the process, and many other factors. However, the decisive factor is the properties of the encapsulated substance.

Physical methods of encapsulation include suspension crosslinking, solvent evaporation, simple and complex coacervation, phase separation, spray drying, fluidized bed spraying, melt crystallization, precipitation, co-extrusion, layering. Thus, the formation of capsules in this case occurs without chemical interaction [11–13].

Chemical methods of microencapsulation include emulsion polymerization, interfacial polymerization, dispersed and interfacial methods [14]. In this case, the shell materials can be monomers, oligomers or polymers having functional groups and capable of participating in reactions of growth or crosslinking of chains with the formation of high molecular weight linear and reticulated polymers.

Chemical methods also include the synthesis of polyelectrolyte nanocapsules, the "Layer-by-layer" method—electrostatic self-assembly, which was proposed by scientists of the Max Planck Institute in 1998 [15,16]. For the first time, the "Layer-by-layer" method was used to form monolayer ultrathin polymer films on a macroscopic substrate. In 1966, the authors [17] proposed to use sequential adsorption to assemble films. In 1991, Decher et al. considered a method for obtaining polyelectrolyte films, which consists in the alternate adsorption of polycations and polyanions on a substrate [18]. The synthesis of polyelectrolyte capsules consists in the deposition of oppositely charged polyelectrolytes on the surface of a solid particle called a template, which can be microparticles of polystyrene, silicon dioxide [19], calcium carbonate [20], cadmium carbonate. The capsule core is most often removed by dissolution. Its material affects the permeability of the capsule shell, its shape, morphology and the rate of removal of the core from the capsule. Calcium carbonate templates are the most frequently used in experiments, since in the case of using silicon dioxide, difficulties arise with the use of hydrofluoric acid, which is used to dissolve it. The use of polystyrene and melamine formaldehyde as templates is limited by their incomplete dissolution. When creating capsules synthesized for medical purposes, calcium carbonate is the most suitable because of its biocompatibility, biodegradability, as well as porous structure and large surface area, which allows it to be used for encapsulating various substances [21].

The formation of polyelectrolyte shells on colloidal particles of different nature is carried out by the method of alternate adsorption of oppositely charged polyelectrolyte macromolecules, as a result of which a shell of various thicknesses can be formed [22–24].

To form the capsule shell, synthetic (polystyrene sulfonate, polyacrylic acid, polydiallyldimethylammonium chloride, etc.) and biocompatible polyelectrolytes (hylauronic acid, sodium alginate, chitosan, L-lysine, etc.) are used. The formation of polyelectrolyte shells occurs mainly through electrostatic interaction; hydrophobic interactions or the formation of hydrogen bonds can also occur [25–27].

The capsule core is most often removed by dissolution. When melamine-formaldehyde latex particles and tetrahydrofuran are used as templates [23,24,28], organic solvents are used.

Modification of microcapsules can be carried out in three ways: through the synthesis of nanoparticles in a polyelectrolyte shell, for example, gold nanoparticles [29–33], by incorporation into the core, or by adsorption of stabilized nanoparticles into a polyelectrolyte shell [34,35]. These methods are currently being developed and improved. The optical and antibacterial properties of silver nanoparticles are of great interest.

The analysis of the cited literature data showed a wide range of works of domestic and foreign scientists aimed at obtaining antibacterial materials for various purposes. Despite the large number of works, the problem of obtaining a stable release form of antibacterial drugs based on silver nanoparticles for finishing textile materials made from natural fibers and implementing the technology for their use remains not fully resolved. There are practically no technologies aimed at creating silver-containing microencapsulated antibacterial agents for processing textile materials for medical purposes.

One of the ways of imparting antimicrobial properties is the development of encapsulated antibacterial drugs and methods of their immobilization on textile materials. Microencapsulation methods make it possible to obtain particles of various sizes—from fractions of a micron to hundreds of microns.

Among such systems, we should especially note polyelectrolyte microcapsules (PEM), a significant property of which is the semi-permeability of the shell, which can allow small molecules to pass through, but retains high-molecular compounds.

This makes it possible to consider PEM as the main method of immobilization of proteins and high-molecular biologically active substances (BAS), when the semipermeable shell of the microcapsule separates the aqueous solution of the substrate from the protein solution and thus protects it from negative external influences. Additional advantages of polyelectrolyte capsules over other similar systems are their mono-dispersity with a wide range of specified sizes; simplicity of regulating their permeability and the possibility of a wide choice of shell material. The shells of such microcapsules can be modified, including various types of ions, functional molecules, nanoparticles.

The most acceptable from a practical point of view are two ways of encapsulating biologically active substances:

-nanoemulsion method;

-synthesis of nanocapsules using templates, where microparticles of calcium carbonate are used to form PEM.

Since the processed textile materials are planned to be used for medical purposes, it is necessary to use biocompatible biodegradable polyelectrolytes to form the capsule shell.

The purpose of this work is the formation of ultrathin multifunctional coatings on cellulose textile material using silver-containing microcapsules formed with the core of a wound-healing compound (oyster mushroom mycelium extract). The coatings are designed to create a cellulose material with high consumer characteristics, capable of inhibiting the vital activity of pathogenic bacteria and fungi, and at the same time exerting a wound-healing effect.

2. Materials and Methods

2.1. Materials

The object of the study was a cotton bleached calico of plain weave with a surface density of 142 ± 7 g/m^2, produced by JSC "Nordtex", Ivanovo.

The following polyelectrolytes were used: chitosan (manufactured by "Bioprogress" Ltd., Shchelkovo, Moscow, Russia) and xanthan gum (manufactured by "INGREDIKO" Ltd., Moscow, Russia).

2.2. Methods

The work uses a set of physical and chemical research methods (atomic absorption spectroscopy, dynamic light scattering, scanning electron microscopy), conventional and original methods for assessing mechanical strength of the fabric and special, including antimicrobial, consumer characteristics of textile materials, generally accepted and original methods for assessing the antimicrobial, antimycotic and consumer characteristics of textile materials. To determine the amount of atomic silver in the capsule shell, an MGA-915 atomic absorption spectrometer (Lumex Instruments Canada, Fraserview Place Mission,

British Columbia, Canada) was used; SEM photographs were taken using a Solver 47 Pro, NT-MDT scanning atomic force microscope (NT-MDT, Zelenograd, Moscow, Russia).

2.3. Silver Nanoparticles Obtaining Procedure

The object of the study was an aqueous solution of silver nitrate $AgNO_3$ of analytical grade, provided by LLC Lenreaktiv, St. Petersburg. The concentration of silver in the investigated solutions varied from 0.24×10^{-7} mol/dm^3 to 0.47×10^{-4} mol/dm^3. An aqueous solution of glyoxal with a concentration of 0.1 to 2 mol/dm^3 was used as a reducing agent, which was prepared by introducing an appropriate sample into cooled bi-distilled water with continuous stirring using a magnetic stirrer. To synthesize silver nanoparticles, a reducing agent solution was added to a silver nitrate solution of a certain concentration. The ratio of the solutions was varied. The prepared solutions of silver nitrate with a reducing agent and a stabilizer were heated to a temperature of 30–90 °C for 5–60 min. The silver reduction reaction was carried out in air.

Determination of the size of silver particles in the studied hydrosols was carried out by dynamic light scattering on a Zetasizer Nano ZS device (Malvern Panalytical, Worcestershire, UK). The size of the synthesized silver nanoparticles used in the study is 2 nm. A sufficiently high uniformity of particle sizes of 96%–100% was achieved.

2.4. Microencapsulation Techniques for Functional Substances
2.4.1. Method of Obtaining Microparticles of Calcium Carbonate

Spherical colloidal particles of $CaCO_3$ were obtained by mixing solutions of $CaCl_2$ and Na_2CO_3 with a concentration of 0.15 M. The reaction mixture was stirred for 5 min under normal conditions. After the completion of the process, $CaCO_3$ particles were washed from Na^+ and Cl^- ions with distilled water and evaporated. Measurements of the size of the obtained calcium carbonate particles and determination of the particle size distribution were performed by laser diffraction on an Analysette 22 NanoTec instrument (Fritsch GmbH, Idar-Oberstein, Germany).

2.4.2. Polyelectrolyte Capsule Shell Synthesis Procedure

To form a polyelectrolyte shell, we used calcium carbonate cores of various diameters and natural polyelectrolytes, for example, negatively charged polyelectrolyte xanthan gum and positively charged chitosan. Calcium carbonate cores have a negative surface charge, so a cationic polyelectrolyte was applied as the first layer. For this purpose, 100 mL of a polyelectrolyte solution with a concentration of 0.001 g/L was added to 0.5 g of cores. The suspension was stirred for 20 min, after which the particles were washed three times with water. The same procedure was carried out using an anionic polyelectrolyte solution. Thereafter, 10 mL of a colloidal solution of silver nanoparticles was added to the system. By the method of alternate adsorption of oppositely charged macromolecules on colloidal particles, a shell consisting of the required number of layers was obtained.

2.4.3. Dissolution of Carbonate Cores Method

The production of hollow polyelectrolyte shells—permeable capsules—was carried out by dissolving $CaCO_3$ cores with the addition of the trisodium salt of ethylenediaminetetraacetic acid (EDTA). As a result, calcium is removed from the capsule due to the formation of a stable complex of this metal with EDTA. A 0.2 M aqueous solution of EDTA with a pH of 7.5 was poured into the capsule suspension and stirred for 20 min, then the suspension was washed three times with distilled water.

2.4.4. Determination of the Sensitivity of Microorganisms to Antimicrobial Drugs by the Disk Method (Diffusion Test)

The method is based on the suppression of the growth of microorganisms on a solid nutrient medium under the action of an antimicrobial drug applied to a paper or tissue disc. As a result of diffusion of the drug into the environment, a concentration gradient of the test drug is formed around the disc—a zone of suppression of the growth of microorganisms. The size of the growth inhibition zone determines the effectiveness of the drug in relation to the studied bacterial culture. Within certain limits, the diameter of the growth inhibition zone is inversely proportional to the minimum inhibitory concentration.

For the study, a suspension containing a standard number of viable cells was used, which was inoculated with a lawn on the surface of a dense nutrient medium—agar Giventalya–Vedminoy, which does not interfere with the diffusion of preparations into Petri dishes. The discs soaked in the test preparations are placed on the inoculation at a distance of 2.5 cm from the center of the dish in a circle. The samples are incubated under conditions favorable for each specific microorganism, in our case 20 h at 35 °C. Then the diameters of growth inhibition zones around the disc were measured in millimeters (taking into account the disc diameter). The area to be measured is the area where the growth of bacteria is completely absent.

2.4.5. Assessment of Wound Healing on Models of Excisional and Burn Wounds

To study the effectiveness of textile materials with microcapsules of biologically active substances and silver nanoparticles on wound healing, samples with four-layer capsules containing silver nanoparticles in various concentrations from 1.75 to 10.10 mg/L were taken. The studies were carried out on outbred rats for 28 days. Before creating excisional and burn skin wounds, the animals were anesthetized. After that, hair was removed on the dorsal side of the body of the rats in the region of the shoulder blades, and the skin was wiped with 70° ethanol. Then one part of the animals received full-thickness wounds with a diameter of 12–14 mm, the other—thermal burns by applying a metal coin heated for 1 min (100 °C) on the skin for 30 s. A bandage in the form of a piece of tissue measuring 2×2 cm^2 was applied to the wounds and fixed from the first day 2 times a day. Immediately after the creation of excisional wounds in some of the animals, the wounds were infected with the Staphylococcus aureus ATCC®6538P™ strain at a dose of 1.5×10^9 cells/mL (10ME). The application of the drug in these animals was started in some rats from the first day, in others from the fifth day.

3. Results

3.1. Protocol for the Synthesis of Biologically Active Substances Nanocapsules of Natural Origin

At this stage of research, a method was developed for encapsulating water-soluble biologically active substances (for example, an oyster mushroom mycelium extract). The direct method of encapsulation was used [35,36], when the injection of a functional substance was carried out during the formation of templates. Spherical colloidal particles of CaCO$_3$ were selected as templates, which were obtained by mixing solutions of CaCl$_2$ and Na$_2$CO$_3$ with a concentration of 0.15–0.33 M. The reaction was expressed by the following equation:

$$CaCl_2 + Na_2CO_3 = CaCO_3 + 2NaCl \tag{1}$$

When the solutions were quickly mixed, an amorphous precipitate of calcium carbonate was formed. After the completion of the process, the CaCO$_3$ templates were washed from Na$^+$ and Cl$^-$ ions with distilled water and evaporated.

To deposit polyelectrolyte layers on particles the method of polyion assembly was used. It was carried out by sequential treatment with oppositely charged polyelectrolytes. To form a polyelectrolyte shell, biodegradable polyelectrolytes were used: chitosan and xanthan gum. Since the calcium carbonate cores have a negative surface charge, a positively charged polyelectrolyte, chitosan, was applied as the first layer. To do this, 100 mL of a 0.1% polyelectrolyte solution prepared in the presence of 0.5 mol/L sodium chloride was added to 1.0 g of the cores. The suspension was stirred for 20 min. After adsorption of each polyelectrolyte layer, the suspension was centrifuged and the particles were washed three times with distilled water. Then the same procedure was carried out using a solution of negatively charged xanthan gum polyelectrolyte, concentration of 0.05%. Further, by the method of alternate adsorption of oppositely charged macromolecules on colloidal particles, it is possible to obtain a shell consisting of the required number of layers. In this work, two- and four-layer capsules were synthesized. At all stages of the experiment, the pH was varied in the range of 6.0–6.2.

Hollow polyelectrolyte shells—permeable capsules—were obtained by dissolving $CaCO_3$ cores with the addition of the trisodium salt of ethylenediaminetetraacetic acid (EDTA). The calcium carbonate from the capsule was removed by forming a stable calcium EDTA complex (see Section 2.4.3).

By varying the conditions of the process (concentration of reagents, temperature, intensity of stirring of the reaction mixture, its duration), it is possible to obtain microspherolites with an average diameter of 2 to 11 μm and a fairly narrow size distribution.

To measure the size of the obtained calcium carbonate particles and determine the particle size distribution, the method of laser diffraction was used on an Analysette 22 NanoTec instrument. With an increase in the stirring time from 0.5 to 5 min, the size of the calcium carbonate particles obtained as a result of reaction (1) decreased from 8.68 to 6.96 μm (Table 1). Stirring the reaction mixture for more than 5 min did not lead to a change in the size of the synthesized particles.

Table 1. Influence of the mixing time of the reaction mixture on the dimensional characteristics of calcium carbonate templates.

Indicators	Value of Indicators						
Stirring time of the reaction mixture, min	0.5	1.5	3	4	5	6	10
Dimensional characteristics of $CaCO_3$ particles, μm	8.68	8.6	7.77	7.68	6.96	6.96	6.96

Oyster mushroom mycelium extract was chosen as a BAS. Basidiomycete Pleurotus ostreatus (oyster mushroom) is widely used in medical practice as producers of various drugs [37]. Extracts obtained from its mycelium are capable of inhibiting the growth of malignant tumors and, at the same time, exhibiting strong immunostimulating activity [38,39]. They are complex systems that include various biologically active components, namely: ubiquitin—like proteins, lectins, proteases, and glucans with multifunctional biomedical activity.

The permeability of the polyelectrolyte shell of a microcapsule can be changed by varying the pH rage [40]. The method is based on shielding the electrostatic interaction between polyelectrolytes, while an increase in the diameter of microcapsules is observed with a sequential change in the pH range from alkaline to acidic. Therefore, for the introduction of biologically active substances (oyster mushroom mycelium extract) inside the microcapsules, it is necessary to acidify the solution. This contributes to the partial "loosening" of the microcapsules, in which pores are created in this case, this allows the BAS to penetrate into the capsule. For this, the system was added 2–5 mL of 3% acetic acid solution to pH = 3.5–4. Then, with stirring, 100 μL of BAS was introduced. After vigorous stirring for 15–20 min, neutralization was carried out with sodium hydroxide solution (1%) to pH = 6. The suspension was centrifuged, the upper fraction of the supernatant was discarded, and distilled water was added.

Based on the developed technique, capsules with di-, tri-, and tetraloid shells, including a BAS solution, were synthesized.

To confirm the presence of capsules in the system under study by dynamic light scattering, the sizes of synthesized capsules were measured on a Photocor Compact-Z device (photon correlation spectroscopy) and photographs were obtained using a Mikmed-6 microscope with a camera.

Table 2 shows various protocols for the preparation of polyelectrolyte shells on calcium carbonate particles.

Table 2. Influence of the capsule formation technique on their appearance and aggregate stability.

No.	Photographs of Particles Obtained Using a Microscope "Micromed 1"	Method for Obtaining Capsules	Particle Size Distribution	
			Particle Size, nm	Percentage, %
1		1.1 1.0 g of calcium carbonate particles were treated for 20 min in a solution of chitosan with a concentration of 0.1%; - rinsing with water; - impregnation with a solution of xanthan gum 0.05% 20 min; - rinsing with water; 1.2. dissolution of the core.	9.7 316.1 7597.0	0.1 24.6 75.3
2		2.1. as in item 1.1 2.2. chitosan layer (according to the technology above); -rinsing with water; 2.3. dissolution of the core; 2.4. acidification and injection of 100 µL BAS; -stirring and neutralization with sodium hydroxide.	496.0 909.7 7989.0	0.4 78.5 21.1
3		The lifetime of the system prepared according to the developed method is 6 days.	522.0 922.8 7998.1	2.0 97.2 0.8

The first sample, as can be seen from the photograph, are hollow capsules, the contents of which are completely transparent. The capsule size was 7000–8000 nm. The capsules presented in the second sample were synthesized according to the developed method. As can be seen from the photograph, the capsule core has a dark color, which corresponds to the color of the BAS injected into the capsule. It can be argued that the technology of forming the shell allows purposefully change of the permeability of the capsule shell by varying the pH of the suspension, thereby ensuring the penetration of biologically active substances into the core. The final size of the capsule had changed to 900–1000 nm (70%), at the same time, large capsules remained, about 8000 nm (20%), and a small number of capsules of 496–522 nm appeared.

The aggregation of the capsules was monitored using an optical microscope. The capsules synthesized according to the developed protocol in the system tend to the formation of associates, however, no associates were observed within a week (Table 2, sample 3). Since the technology involves the application of freshly prepared microencapsulated compositions to the textile material, the stability of the suspension for 6 days is sufficient. Using a scanning electron microscope, photographs of microcapsules applied to the textile material were obtained (Figure 1).

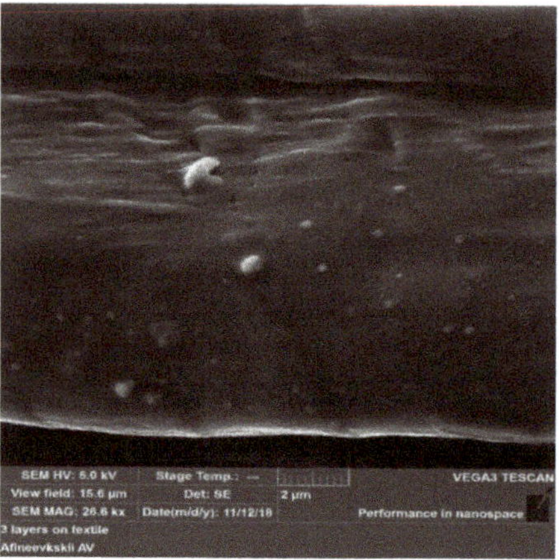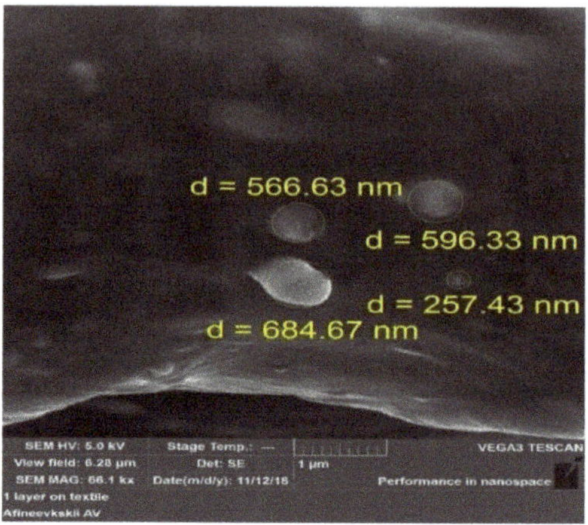

Figure 1. SEM photograph of textile material containing polyelectrolyte microcapsules.

3.2. Atomic Adsorption Microscopy of Microcapsules Containing Silver Nanoparticles

To provide the capsule shell bactericidal properties, a technology was developed for immobilizing silver nanoparticles into its composition. The idea was that the preparations applied to the textile material first showed a bactericidal effect, disinfecting the wound, and then the released medicinal substance had a directed effect. For this, a positive polyelectrolyte, chitosan, was applied to the calcium carbonate core in the first layer, washed three times with distilled water, and a layer of negative polyelectrolyte, xanthan gum, was applied in the second layer and washed with water. Before applying the third polyelectrolyte layer consisting of positively charged chitosan, microparticles were impregnated in a solution of silver nanoparticles.

The use of atomic absorption spectroscopy made it possible to determine the presence of silver nanoparticles in the microcapsules. It was shown that the amount of silver in the nanoform per gram of microcapsules was 0.0067 mg.

The resulting capsules were applied to the textile material according to the following process: impregnation with a solution of chitosan with a concentration of 5 g/L, drying at 1000 °C for 3 min, impregnation with a nanodispersion of the encapsulated preparation, drying at a temperature of 100 °C for 3 min, after which the antibacterial activity of the samples was determined in relation to gram positive and gram-negative microorganisms and fungi (Table 3).

The obtained results confirmed the antibacterial efficiency of the developed chemical product.

3.3. Functional Characteristics of Finished Textile Materials

The hydrophilic properties of the treated samples of textile materials were monitored by such indicators as capillarity, wettability, and water absorption, determined in accordance with standard methods [41]. It was found that before and after the application of the antibacterial coating, the capillarity remains at the level of 150–155 mm, the wettability is less than 1 s, the water absorption is −20–20.2 g/m^2, which fully meets the requirements for these materials.

The reparative effect of textile materials with a multifunctional coating was investigated on outbred rats for 28 days according to the following criteria:

- local inflammatory response;
- results of planimetric studies;
- timing of wound healing;

- determination of the index of acceleration of healing in the experimental groups in relation to self-healing wounds in the control groups.

Table 3. Antibacterial activity of the obtained samples in relation to Staphylococcus aureus, Escherichia coli, and Candida albicans.

Microorganism	Photo of Results/Zone of Delay, mm	
	With Silver	**Without Silver**
Staphylococcus aureus—a spherical, immobile, aerobic (airborne) bacterium, positively stained according to Gram, which causes various diseases in children and adults.	8 mm	0 mm
Escherichia coli—a type of gram-negative rod-shaped bacteria, widespread in the lower intestines of warm-blooded animals. However, among *E. coli*, there are also species that can cause various infectious diseases in humans, ranging from common intestinal disorders to sepsis.	3 mm	0 mm
Candida albicans is a diploid fungus, pathogen, anaerobic microorganism. It may be a consequence and an etiological factor of the occurrence of multiple human infections.	13 mm	2 mm

During the experiments, all animal welfare standards were observed

Evaluation according to these criteria was carried out on 4, 8, 11, 15, 18, and 22 days after the start of treatment.

All wounds healed the same way. No inflammatory processes were registered. Planimetric survey data are presented in Table 4.

The results obtained showed that only materials with microencapsulated drug concentrations of more than 5.93 mg/L had a positive effect on the rate of healing of burn wounds. Materialss with any of the suggested concentrations had a positive effect on excision wounds, up to 17% on uninfected ones, and up to 7.5% on infected ones.

Application of the Staphylococcus aureus ATCC 6538P strain to wounds at a dose of 1.5×109 cells/mL (10ME) did not lead to wound infection (except for one). However, this method of creating purulent wounds, described in the literature, requires improvement.

The result of the reparative action of textile materials with biologically active substances on such wounds is ambiguous.

Table 4. Results of planimetric studies of wounds.

Model	The Beginning of Wound Treatment	Drug Concentration, mg/L	Number of Animals	Percentage of Healing Relative to the Primary Wound,%						Healing Acceleration Index,%
		Day from Model Creation		4	8	11	15	18	22	-
Burn wound	From the first day	-	1	0	0	0	23.1	38.5	53.9	-
		1.75	1	0	0	0	7.6	30.8	46.2	−14.3
		3.84	1	0	0	0	0	30.8	46.2	−14.3
		5.93	2	0	0	0	11.6	29.7	44.6	−17.3
		8.01	2	0	0	0	14.3	27.4	56.6	+5.0
		10.1	2	3.6	3.6	21.4	26.2	42	57.1	+5.9
Excision wound	From the first day	-	1	0	35.7	57.1	85.7	85.7	85.7	-
		1.75	1	0	33.3	50.0	66.7	75,0	83.3	−2.8
		3.84	1	0	41.7	66.7	83.3	91.7	91.7	+7.0
		5.93	2	3.6	47.8	64.0	73.1	77.6	92.3	+7.7
		8.01	2	0	28.7	48.7	74.3	87.6	100	+16.7
		10.1	2	0	45.0	55.8	78.3	90.0	100	+16.7
Excision wound + staphylococcus	From the first day	-	2	0	46.5	60.7	75.0	92.9	92.9	-
		1.75	1	7.1	42.9	71.4	78.6	85.7	92.9	0
		3.84	2	9.8	46.9	63.4	89.8	93.2	96.9	+4.3
		5.93	2	0	30.8	61.4	80.8	84.6	92.3	−0.6
		8.01	2	0	37.5	62.6	93.8	100	100	+7.6
	From the fifth day	1.75	1	0	30.8	38.5	69.2	84.6	92.3	−0.7
		3.84	1	0	40.0	60.0	66.7	100	100	+7.6
		5.93	2	0	37.3	61.4	92.3	92.3	96.2	+3.6
		8.01	2	0	25.0	31.3	43.8	56.3	75.1	−19.2

4. Conclusions

A technology for the preparation of capsules by the method of sequential adsorption of polyelectrolytes of chitosan and xanthan gum, the shells of which included silver nanoparticles, and the core was a wound-healing agent was developed. The capsule shell synthesized using a polyelectrolyte assembly consisted of biodegradable and safe natural polymers. Nano-silver located between the polyelectrolyte layers gave the finish an antimicrobial effect. In this study, oyster mushroom mycelium was successfully used as a model filling of the capsule core. The proposed model of the formation of microcapsules, the shells of which are doped with silver nanoparticles, allows the most targeted effect on the focus of infection, providing an antimicrobial effect. The encapsulated preparation of the oyster mushroom mycelium extract was released for a long time and provided in this case a more effective wound healing effect.

Cellulose tissue samples coated with silver-containing microcapsules have demonstrated high antibacterial, antimycotic activity and wound healing properties.

It was found that the optimal concentration of the microencapsulated drug for effective wound healing is 1%–2% of the weight of the textile material. In the case of burn wounds, to restore the skin, the material should be changed at least after two weeks.

Changing the filler of the capsule core (biologically active substance) makes it possible to impart a spectrum of different properties to the developed drug, and to the textile material—various types of final finishing.

Author Contributions: conceptualization, methodology, writing—preparation of the initial draft, writing—review and editing, O.O. and O.K.; data curation, research, visualization L.P. and A.L.; software, investigation, visualization, S.S. and E.V. All authors have read and agreed to the published version of the manuscript

Funding: This work was financially under the agreement with LLC "Smart Textile" No. 09.121.18., Carried out under the HealthNet program of the Innovation Promotion Fund.

Institutional Review Board Statement: The ethical review and approval of this study was rejected due to the lack of a need for this procedure in the Russian Federation. There is no independent law on the protection of experimental animals. Experiments on living organisms are regulated by order of the USSR Ministry of Health No. 755 of August 12, 1977 "On measures to further improve organizational forms of work using experimental animals." All experiments were carried out according to the above order.

Informed Consent Statement: Informed consent was obtained from all subjects involved in the study.

Data Availability Statement: The data presented in this study are available on request from the corresponding author. The data is not publicly available due to the lack of patents. In the near future, patents will be issued and obtained, and data will become publicly available.

Acknowledgments: The authors would like to express their gratitude to Zacharov S.V. for his help in the microbiological studies and assessment of the healing properties of the samples (LLC Inbiopharm). The work was done on the equipment of the Center for Collective Use of Ivanovo State University of Chemistry and Technology.

Conflicts of Interest: The authors declare no conflict of interest.

References

1. Pinto, R.J.; Marques, P.A.; Neto, C.P.; Trindade, T.; Daina, S.; Sadocco, P. Antibacterial activity of nanocomposites of silver and bacterial or vegetable cellulosic fibers. *Acta Biomater.* **2009**, *5*, 2279–2289. [CrossRef] [PubMed]
2. Krzywicka, A.; Megiel, E. Silver-Polystyrene (Ag/PS) Nanocomposites Doped with Polyvinyl Alcohol (PVA)—Fabrication and Bactericidal Activity. *Nanomaterials* **2020**, *10*, 2245. [CrossRef]
3. Valdés, A.; Ramos, M.; Beltran, A.; Garrigos, M.C. Recent Trends in Microencapsulation for Smart and Active Innovative Textile Products. *Curr. Org. Chem.* **2018**, *22*, 1237–1248. [CrossRef]
4. Ghost, S.K. *Functional Coatings by Polymer Microencapsulation*; Wiley-VCH: Weinheim, Germany, 2006; p. 378.
5. Krolevets, A.A.; Tyrsin, Y.A.; Bykovskaya, E.E. Application of nano- and microencapsulation in pharmaceuticals and food industry. *Vestnic Rossiiskoy Akademii Estestvennykh Nauk* **2013**, *1*, 77–84.
6. Cheng, S.Y.; Yuen, C.W.M.; Kan, C.W.; Cheuk, K.K.L. Development of cosmetic textiles using microencapsulation technology. *Res. J. Text. Appar.* **2008**, *12*, 41–51. [CrossRef]
7. Sarma, S.J.; Pakshirajan, K.; Mahanty, B. Chitosan-coated Alginate–polyvinyl Alcohol Beads for Encapsulation of Silicone Oil Containing Pyrene: A Novel Method for Biodegradation of Polycyclic Aromatic Hydrocarbons. *J. Chem. Technol. Biotechnol.* **2011**, *2*, 266–272. [CrossRef]
8. Kumar, B.P.; Chandiran, I.S.; Bhavya, B.; Sindhuri, M. Microparticulate Drug Delivery System: A Review. *Indian J. Pharm. Sci. Res.* **2011**, *1*, 19–37.
9. Borodina, T.N. Obtaining and Research of Biodegradable Polyelectrolyte Microcapsules with Controlled Release of Proteins, DNA and Other Bioactive Compounds. Ph.D thesis, Russian Academy of Sciences, Moscow, Russia, 16 May 2008; p. 119.
10. Anal, A.K.; Singh, H. Recent Advances in Microencapsulation of Probiotics for Industrial Applications and Targeted Delivery. *Trends Food Sci. Technol* **2007**, *18*, 240–251. [CrossRef]
11. Sanjoy, D. Microencapsulation techniques and its practices. *Int. J. Pharm. Sci. Technol.* **2011**, *6*, 1–23.
12. Silva, P.T.D.; Fries, L.L.M.; Menezes, C.R.D.; Holkem, A.T.; Schwan, C.L.; Wigmann, É.F.; Silva, C.D.B.D. Microencapsulation: Concepts, mechanisms, methods and some applications in food technology. *Ciênc. Rural* **2014**, *44*, 1304–1311. [CrossRef]
13. Salaün, F. *Microencapsulation Technology for Smart Textile Coatings*; Active Coatings for Smart Textiles; Woodhead Publishing: Southston, UK, 2016; pp. 179–220.
14. Patel, K.R.; Patel, M.R.; Mehta, T.J.; Patel, A.D.; Patel, N.M. Microencapsulation: Review on novel approaches. *Int. J. Pharm.* **2011**, *648*, 894–911.
15. Donath, E.; Sukhorukov, G.B.; Caruso, F.; Davis, S.A.; Möhwald, H. Novel hollow polymer shells by colloid-templated assembly of polyelectrolytes. *Angew. Chem. Int. Ed.* **1998**, *37*, 2201–2205. [CrossRef]
16. Sukhorukov, G.B.; Donath, E.; Lichtenfeld, H.; Knippel, E.; Knippel, M.; Budde, A.; Möhwald, H. Layer-by-layer self assembly of polyelectrolytes on colloidal particles. *Colloids Surf. A: Physicochem. Eng. Asp.* **1998**, *137*, 253–266. [CrossRef]
17. Iler, R.K. Multilayers of colloidal particles. *J. Colloid Interface Sci.* **1966**, *21*, 569–594. [CrossRef]

18. Decher, G.J.D.H.; Hong, J.D. Buildup of ultrathin multilayer films by a self-assembly process: II. consecutive adsorption of anionic and cationic bipolar amphiphiles on charged surfaces. *Macromol. Chem., Macromol. Symp.* **1991**, *46*, 321–327. [CrossRef]
19. Schuetz, P.; Caruso, F. Copper-assisted weak polyelectrolyte multilayer formation on microspheres and subsequent film crosslinking. *Adv. Funct. Mater.* **2003**, *13*, 929–937. [CrossRef]
20. Volodkin, D.V.; Petrov, A.I.; Prevot, M.; Sukhorukov, G.B. Matrix polyelectrolyte microcapsules: New system for macromolecule encapsulation. *Langmuir* **2004**, *20*, 3398–3406. [CrossRef]
21. Combes, C.; Bareille, R.; Rey, C. Calcium carbonate–calcium phosphate mixed cement compositions for bone reconstruction. *J. Biomed. Mater. Res. Part A* **2006**, *79*, 318–328. [CrossRef]
22. Aisina, R.B.; Kazanskaya, N.F. Microencapsulation of physiologically active substances and their application in medicine. Results of science and technology. *Biotechnol. Ser.* **1986**, *6*, 6–52.
23. Lameiro, M.H.; Lopes, A.; Martins, L.O.; Alves, P.M.; Melo, E. Incorporation of a model protein into chitosan–bile salt microparticles. *Int. J. Pharm.* **2006**, *312*, 119–130. [CrossRef]
24. Grenha, A.; Seijo, B.; Remunán-López, C. Microencapsulated chitosan nanoparticles for lung protein delivery. *Eur. J. Pharm. Sci.* **2005**, *25*, 427–437. [CrossRef] [PubMed]
25. Kozlovskaya, V.; Kharlampieva, E.; Drachuk, I.; Cheng, D.; Tsukruk, V.V. Responsive microcapsule reactors based on hydrogen-bonded tannic acid layer-by-layer assemblies. *Soft Matter* **2010**, *6*, 3596–3608. [CrossRef]
26. Lee, D.; Rubner, M.F.; Cohen, R.E. Formation of nanoparticle-loaded microcapsules based on hydrogen-bonded multilayers. *Chem. Mater.* **2005**, *17*, 1099–1105. [CrossRef]
27. Such, G.K.; Johnston, A.P.R.; Caruso, F. Engineered hydrogen-bonded polymer multilayers: From assembly to biomedical applications. *Chem. Soc. Rev.* **2010**, *40*, 19–29. [CrossRef] [PubMed]
28. Lambert, G.; Fattal, E.; Couvreur, P. Nanoparticulate systems for the delivery of antisense oligonucleotides. *Adv. Drug Deliv. Rev.* **2001**, *47*, 99–112. [CrossRef]
29. Parakhonskiy, B.V.; Bedard, M.F.; Bukreeva, T.V.; Sukhorukov, G.B.; Mohwald, H.; Skirtach, A.G. Nanoparticles on polyelectrolytes at low concentration: Controlling concentration and size. *J. Phys. Chem. C* **2010**, *114*, 1996–2002. [CrossRef]
30. Antipov, A.A.; Sukhorukov, G.B.; Fedutik, Y.A.; Hartmann, J.; Giersig, M.; Möhwald, H. Fabrication of a novel type of metallized colloids and hollow capsules. *Langmuir* **2002**, *18*, 6687–6693. [CrossRef]
31. De Geest, B.G.; Skirtach, A.G.; De Beer, T.R.; Sukhorukov, G.B.; Bracke, L.; Baeyens, W.R.; De Smedt, S.C. Stimuli-Responsive Multilayered Hybrid Nanoparticle/Polyelectrolyte Capsules. *Macromol. Rapid Commun.* **2007**, *28*, 88–95. [CrossRef]
32. Bagaria, H.G.; Kadali, S.B.; Wong, M.S. Shell thickness control of nanoparticle/polymer assembled microcapsules. *Chem. Mater.* **2010**, *23*, 301–308. [CrossRef]
33. Yuan, W.; Lu, Z.; Li, C.M. Controllably layer-by-layer self-assembled polyelectrolytes/nanoparticle blend hollow capsules and their unique properties. *J. Mater. Chem.* **2011**, *21*, 5148–5155. [CrossRef]
34. Radziuk, D.; Skirtach, A.; Sukhorukov, G.; Shchukin, D.; Möhwald, H. Stabilization of silver nanoparticles by polyelectrolytes and poly (ethylene glycol). *Macromol. Rapid Commun.* **2007**, *28*, 848–855. [CrossRef]
35. Skirtach, A.G.; Antipov, A.A.; Shchukin, D.G.; Sukhorukov, G.B. Remote activation of capsules containing Ag nanoparticles and IR dye by laser light. *Langmuir* **2004**, *20*, 6988–6992. [CrossRef] [PubMed]
36. Déjugnat, C.; Sukhorukov, G.B. pH-Responsive Properties of Hollow Polyelectrolyte Microcapsules Templated on Various Cores. *Langmuir* **2004**, *20*, 7265–7269. [CrossRef] [PubMed]
37. Zaikina, N.A.; Kovalenko, A.E.; Galynkin, V.A.; Dyakov, Y.T.; Tishenkov, A.D. *Fundamentals of Biotechnology of Higher Fungi: Textbook*; Prospekt Nauki: St. Peterburg, Russia, 2007.
38. Korsun, V.F.; Krasnopolskaya, L.M.; Korsun, E.V.; Avkhukov, M.A. *Antitumor Properties of Fungi*; Mailer: Moscow, Russia, 2012.
39. Gerasimenya, V.P.; Zakharov, S.V.; Putyrsky, L.A. *Antineoplastic Action of Oyster Mushroom Mycelium Extract. Experimental and Clinical Studies: Information Materials*; Issue 3; Inbiopharm LLC: Moscow, Russia, 2009.
40. Mauser, T.; Déjugnat, C.; Sukhorukov, G.B. Reversible pH-dependent properties of multilayer microcapsules made of weak polyelectrolytes. *Macromol. Rapid Commun.* **2004**, *25*, 1781–1785. [CrossRef]
41. GOST 3816-81 (ISO 811-81) Textile Fabrics. *Methods for Determining Hygroscopic and Water-Repellent Properties*; IPK Publishing House of Standards: Moscow, Russia, 1998.

Article

Multifunctional Polymer Coatings of Fusible Interlinings for Sewing Products

Nadezhda Kornilova [1,*], **Albina Bikbulatova** [2,*], **Sergey Koksharov** [3], **Svetlana Aleeva** [3], **Olga Radchenko** [1] **and Elena Nikiforova** [1]

[1] Engineering Center for Textile and Light Industry, Ivanovo State Polytechnic University, Sheremetevsky Ave. 21, 153000 Ivanovo, Russia; sva@isc-ras.ru (O.R.); nikiforova@ivgpu.com (E.N.)
[2] Institute of Industrial Engineering, Information Technology and Mechatronics, Moscow State University of Food Production, Volokolamskoe Highway, 11, 125080 Moscow, Russia
[3] Laboratory of Chemistry and Technology of Modified Fibrous Materials, G.A. Krestov Institute of Solution Chemistry of the Russian Academy of Sciences, Akademicheskaya st, 1, 153045 Ivanovo, Russia; nkorn@ivgpu.com (S.K.); bikbulatovaaa@mgupp.ru (S.A.)
* Correspondence: nkorn@mail.ru (N.K.); albina-bikbulatova@yandex.ru (A.B.); Tel.: +7-905-107-69-89 (N.K.); +7-919-727-41-80 (A.B.)

Abstract: The aim of the study was to improve the range of adhesive interlining materials for clothes to add to the product a complex of improved consumer (shape stability, wear resistance) and special (protective, health-improving) properties. We have found that the modification of the adhesive coating of interlining materials with oligoacrylates provides a transition from the traditional discrete 2D glue interlayers between the bonded materials formation to the highly branched 3D structures of the interfacial layer of composites creation. Using dynamic light scattering method, IR spectroscopy, differential scanning calorimetry, gas adsorption, optical and scanning probe microscopy, methods of textile materials science, we have identified technological approaches that ensure the polymer coating penetration into the intrafiber nanopore spaces of the textile layers. We have identified ways to control the elastic-deformation properties of composites in the modifying the adhesive interlining material process in order to design its properties for the requirements of various garment models. We found a unique possibility of using graftable oligoacrylate dispersion to stabilize the nanodispersed state of functional fillers (silica, nanoferrites) and increase the uniformity of their distribution in the composite structure.

Keywords: nanostructured polymer coating; polyacrylate dispersion; nanodispersed fillers; graft copolymers; composite parts of a garment

Citation: Kornilova, N.; Bikbulatova, A.; Koksharov, S.; Aleeva, S.; Radchenko, O.; Nikiforova, E. Multifunctional Polymer Coatings of Fusible Interlinings for Sewing Products. *Coatings* **2021**, *11*, 616. https://doi.org/10.3390/coatings11060616

Academic Editor: Alessandro Pezzella

Received: 23 April 2021
Accepted: 19 May 2021
Published: 21 May 2021

Publisher's Note: MDPI stays neutral with regard to jurisdictional claims in published maps and institutional affiliations.

1. Introduction

Fusible interlining materials (FIM) are used in production of a wide variety of garments: men's suits and shirts, women's jackets, skirts and trousers, special-purpose clothes for protection from various weather- and workplace-related effects, etc. The main purpose of applying these materials is to properly shape a garment and ensure shape stability in wear [1,2].

FIM are textile materials coated with a thermoplastic polymer (adhesive) in the form of a powder or paste in such a way that it remains on the material surface in the form of dots [3]. The mechanism of action of thermoplastic adhesives consists in polymer melt interaction with a fibrous base. This ensures strong bonding of fabric layers and makes the fused fabric composite rigid by forming a polymer matrix between the layers. The process of manufacturing a garment with good shape stability properties consists of three consecutive stages:

- making a fabric composite by gluing flat pieces of the shell fabric (main material) and FIM together (fusing);

- stitching the garment pieces with together;
- shaping the finished composite garment during wet-heat treatment (WHT).

Fusible interlining materials differ from each other in the fabric base structure, nature of the thermally fusible polymer, its amount and deposition method. The rigidity of a composite normally depends on the fiber composition of the fabric base, its surface density, mutual orientation of the warp threads of the shell and interfacing, type and amount of the thermally fusible polymer [3–5]. If the size of adhesive dots and their density increase, it results in effective growth in the fused fabric composite rigidity [2,5,6]. The higher rigidity makes it more difficult to shape the garment, and higher glue content leads to deterioration of hygienic properties, such as breathability.

The properties of FIM are selected considering the designed garment shape and the type of the main material (MM). The more rigid the garment form and the softer the MM, the higher bending deformation resistance fused fabric composite must have [7]. For example, the rigidity value (EI) of a fused fabric composite in different parts of a men's suit flap is required to range $(4.1–39.8) \times 10^{-3}$ N·cm^2. In small volume soft-plastic silhouette form garment models, the maximum EI values within one garment fabric piece can be 2.7 times higher than the minimum ones, whereas in the most rigid models of large volume, this difference equals 5.4 times [5]. In practice, shape stability is achieved by varying the number of interlining layers in different parts of a garment piece using several types of FIM with different rigidity and elasticity degrees. Among the disadvantages of this approach are the thick inner fabric layer that makes the garment heavier, the unpredictable deformation behavior of the garment in wear, poorer hygroscopicity and air permeability caused by the increase in the number of FIM layers and mass of the thermoplastic polymer. Additionally, it is quite natural that the very process of manufacturing of a composite garment piece increases its rigidity and lowers its shaping ability at the first production stage (fusing), which makes it more difficult to put the garment into the required shape at the third stage (final WHT).

An alternative to the use of FIM is the method of direct deposition of viscous polymer compositions in the form of a certain pattern onto the backside of the main material [3,8]. In such case, a polymer matrix is formed in the MM structure. Among the unquestionable advantages of the technique are the possibility of gradient reinforcement of a garment piece in order to regulate the properties of its separate parts, retention of the shaping ability by the composite fabric piece until the finishing WHT stage, and reduction in the garment weight. The main drawback is the risk of polymer penetration through the main material and spoiling of its outside surface appearance.

To ensure that the garment parts look the same, it is reasonable to introduce gradient changes in the stress-strain properties of the polymer matrix formed within the MM + FIM composite structure. This can be done through zonal deposition of a special polymer coating on the FIM surface. When selecting the coating for FIM surface modification, it is reasonable to employ the latest achievements in the field of synthesis of polymers with a complex spatial architecture. Quite promising are also methods of obtaining polymer matrix and interfacial layer in the form of molecular brushes and comb-like structures with numerous side chains attached to the backbone, which makes the macromolecule rigidity tens of times higher [9,10].

It is possible to additionally regulate the stress-strain properties of composite materials by using nanodispersed fillers with a high elasticity modulus, for example, silicon dioxide (SiO$_2$). One of the problems of polymer material modification by nanoparticles is related to their aggregation and nonuniform distribution within the polymer bulk [11,12]. Solving this problem will make it possible to provide fused fabric composites with additional properties by changing the filler type. For example, garments that have pieces with highly coercive nanoparticles in the fibrous base structure can have a positive effect on human health: improve the adaptation and regeneration capacity in stressful situations and protect the human body from high temperatures, acoustic, ultrasonic and electromagnetic exposure, impulsive loads and vibrations, as well as chemical and biological factors [13,14].

We have proposed a technology of FIM modification using a special polymer coating based on selecting a modifying polymer (MP) dispersion that can interact with a thermoplastic polymer (TP) forming a highly branched graft-copolymer, with the lateral branches penetrating into the pore system of the fibrous materials [15,16]. The conditions of producing copolymers of this type within the structure of the fused fabric composite formed must be suitable for MP and TP interaction and retention of the adhesive properties for TP and penetration into the fibrous component pores for MP. The formation of a highly branched polymer layer structure must not lead to an additional increase in the composite material rigidity before the garment takes its final shape [17]. The MP aqueous dispersion consistency must correspond to the conditions of screen printing on the FIM surface, and the MP dispersion degree must be high enough for the particles to diffuse into the inner volume of the fiber during drying process [18].

This paper presents the results of studying the possibility of regulating the stress-strain and consumer properties of composite garment parts by changing the conditions of polymer coating modification in order to obtain highly branched 3D interface structures. Special attention is paid to the degree of the modifying dispersion penetration into the fibrous material and introduction of nanodispersed fillers into the polymer matrix.

2. Materials and Methods

Five types of suite fabrics were used as the main materials in the study (Table 1).

Table 1. Characteristics of the main materials.

Symbol	Fibrous Composition (%)	Surface Density $M_S{}^{MM}$ (g/m^2)	Stiffness EI$_{MM}$, 10^{-3} (N·cm^2)		Shaping Ability A_{MM} (%)		Air Permeability Q_{MM} (dm^3/s m^2)
			Warp	Weft	Warp	Weft	
MM1	viscose 55, wool 35, polyester 10	240 ± 3	4.17	3.11	30.3	33.1	183
MM2	viscose 50, polyester 50	185 ± 2	2.3	2.36	35.3	34.2	275
MM3	viscose 80, polyester 20	190 ± 2	2.0	2.0	36.1	36.1	261
MM4	wool 53, polyester 44, elastane 3	167 ± 3	2.76	2.29	34.0	34.9	248
MM5	wool 99, elastane 1	167 ± 3	5.6	3.3	26.4	33.5	218

For modification, we used standard FIM, based on weft knit with polyamide adhesive dots on one side of it (Table 2).

Table 2. Characteristics of standard FIM.

Symbol	Manufacturer	Fibrous Composition (%)	Surface Density M_S (g/m^2)	Weft Threads Mass Fraction G_{WT} (%)	Adhesive Coating Area [1], S_{TP} (%)
FIM1	Shanghai Uneed Textile Co.,Ltd, Shanghai, China	polyester 30, viscose70	80	60.8	13.2
FIM2	Shanghai Uneed Textile Co.,Ltd, Shanghai, China	polyester 30, viscose 70	65	54.4	16.7
FIM3	Kufner Textile Group, Unterhaching, Germany	polyester 27, viscose 73	58	70.5	18.7
FIM4	Iskozh JSC, Neftekamsk, Russia	polyester 60, cotton 40	75	60.8	25.2
FIM5	Iskozh JSC, Neftekamsk, Russia	polyester 100	70	-	15.8

[1] The adhesive coating area was determined by the number of adhesive dots in 1 cm^2 of FIM multiplied by the average area of one adhesive dot and divided by 100.

Samples of aqueous oligoacrylate dispersions Akremos (LLC "Pilot Plant of Acrylic Dispersions", Dzerzhinsk, Russia), Akratam AS and Anzal ("Pigment" Ltd., Tambov, Russia) with the nonvolatile substance content from 30 to 50 wt.% were used as the MP.

Two types of mechanical effects were applied to achieve particle disaggregation in the hydrosols: ultrasound treatment in a UZDN-2T disperser (LLC "U-RosPribor", Belgorod, Russia) at a frequency of 22 kHz and a combination of high shear stress, ultrasound and cavitation on a rotary-pulse activator (RPA) at the shear rate of $(0.5–17.4) \times 10^4$ s^{-1}.

Hydrosols of detonation nanodiamonds (DND) (Ioffe Physico-Technical Institute, St. Petersburg, Russia) and colloidal silicon dioxide (SD) acted as the nanodispersed filler (Guangzhou Jiechuang Trading Co. Ltd., Guangzhou, China; SiO_2 content 25%–26%, purity 99.5%). The DND was prepared by the oxidative synthesis method and had a zeta-potential of -50 mV and solid phase content of 3.62 wt.%.

The two-component MP systems with DND or SD were obtained by mixing 100 mL of an ultrasonically dispersed filler with 10 mL of a mechanically activated MP dispersion or by treating the mixture of the initial compounds with the required ratio of components in the RPA.

The size of the hydrosol nanoparticles was measured by the dynamic light scattering method on a Zetasizer Nano ZS analyzer (Malvern Panalytical, Malvern, UK); the signal accumulation time in a series of three measurements was 20 min. The analysis of the measurement results was carried out by an automated program based on the solution of the Fredholm integral equation of the first kind with an exponential kernel for the normalized correlation function [19]. To increase the recording ability of the measuring system, taking into account the recommendations [20] for the study of polyfraction systems in the results processing window, the value of the Lower Threshold indicator "0.05" must be corrected to "0" in accordance with the recommendations for studying polyfractional systems.

The MP interaction with the thermally fusible polymer was evaluated by the methods of IR-spectroscopy on an AVATAR 360 FT-IR ESP Fourier transform IR spectrometer (SpectraLab Scientific Inc., Markham, ON, Canada) and differential scanning calorimetry on a DSC 204 F1 Phoenix apparatus (NETZSCH, Selb, Germany) equipped with a μ-sensor. The two-component objects were prepared by introducing 10 wt.% of a powdered thermally fusible polymer into an MP hydrosol, then casting the samples on glass templates and drying them in air.

To develop the microporous structure of the FIM5 textile base polyester fiber, we carried out surface saponification in a boiling NaOH solution (0.25–1 mol/L) in the presence of Alcamon OS-3—a quaternary ammonium compound (0.3 g/L). The processing duration was varied from 3 to 10 min and was followed by washing with cold and hot (80 °C) flow-through water until the phenolphthalein reaction became neutral. The effectiveness of the fiber surface modification was evaluated by the low-temperature nitrogen adsorption-desorption method at 77 K on a NOVA 1200e gas sorption analyzer (Quantachrome, Boynton Beach, FL, USA) to find out the material porosity (VP, m^3/g) and inner pore size distribution.

The area of the FIM surface covered with the modifying dispersion—SA—was varied within the range from 0.35 to 0.65 by changing the screen patterns that had different density and line thickness values. The patterns of MP dispersion deposition and the respective values of the relative interlining reinforcement area are shown in Table 3.

Table 3. Characteristics of patterns for applying MP.

Number	1	2	3	4	5
Pattern of MP dispersion deposition	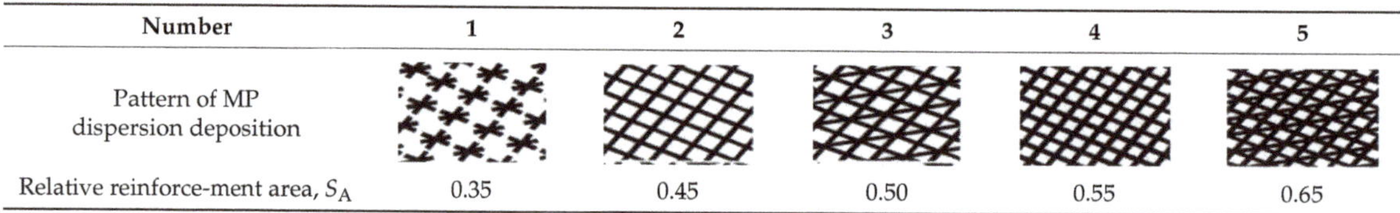				
Relative reinforce-ment area, S_A	0.35	0.45	0.50	0.55	0.65

The fabric composites were made by fusing the main material with a standard or modified FIM on a Japsew SR-600 press (Japsew Corporation, Shanghai, China) a temperature of 110 °C for 20 s. The wet-heat treatment of the fused fabric composites was carried out on a Malkan UPP1AVK (Malkan Machinery, İstanbul, Turkey) press at a temperature of 140 °C for 30 s. The wetting reached 20%–30%.

The processing characteristics and consumer properties of the fused fabrics and finished composites with standard and modified FIM were estimated by the following indicators: thickness, stiffness, elasticity, shaping ability, bond strength, hygroscopicity, air permeability, and shape stability factor.

The indicator thickness was conducted in accordance with the ISO 5084:1996 standard.

The indicator stiffness (EI) of textile liner and fused panel was calculated in accordance with GOST RF 10550-93 using contactless console methodology. The strips of material size of 160×30 mm^2 were used. They were placed horizontally on the top side of the machine and pressed by the load size of 20 mm in the middle. After that, the sides of anchor were dropped down, the ends of material hanged loose due to gravitational force. After a minute, the amount of overhang of the strips ends (f) was measured. The value of stiffness EI (10^{-3} N·cm^2) was calculated using following equation:

$$EI = \frac{42.046 \times m}{0.5755 \times f^3 - 2.411 \times f^2 + 8.502 \times f}$$

where m is the mass of material strip in grams.

The indicator elasticity (U) was calculated using the ring method in accordance with GOST RF 10550-93 using ring methodology. A sample 95×20 mm^2 was clamped on the removable site. The top surface of the sample was outside. The sample should take the form of correct ring. Then it was loaded up to the sample sagging was 1/3 of diameter ($S_0 = 30$ mm) during 30 s. After load breaking the sample laid off during 30 s and the residual deflection S_1 was measured. The value of indicator U (%) was calculated using following equation:

$$U = \frac{S_0 - S_1}{S_0 \times 100\%}$$

The indicator shaping ability (A) characterizes the relative size of the material (composite) field which repeats the three-dimensional surface. It was conducted in accordance with patent RF No.234347 "The method of measurement molding ability of textile liner". For determination of molding ability, the sphere with diameter 150 mm was set on the tripod, which was fixed on the stand. The sample of material the size 350×350 mm^2 was placed from above the sphere. For ensuring a fit of the material to the sphere, folds of the same depth were made in the warp and weft directions. The indicator "shaping ability" (A) was calculated using following equation:

$$A = \frac{\alpha}{\alpha_{max}}$$

where α is a central angle of recurrence by material (composite) the surface of sphere, degrees; α_{max} is a maximum possible angle of recurrence ($\alpha_{max} = 1800$).

Methods for determining indicators EI, U, A are detailed in the article [17].

The indicator bonding strength (P) was conducted in accordance with GOST USSR 28832-90; sample preparation and conducting tests were realized in accordance with the ASTM D 2724 standard under the following conditions: sample size was 150×30 mm^2, size of unglued piece was 40 mm, the distance between clamps was 50 mm, the speed of moving motile clamp was 100 mm/min. The indicator bonding strength (P, N/10 cm) was calculated using equation:

$$P = \frac{F}{300}$$

where F is average value of the force required to separate the interlining and shell fabric.

Indicator hygroscopicity was conducted in accordance with ISO 811-81 Textile fabrics, Methods for determination of hygroscopic and water-repellent properties.

Indicator air permeability (Q) was conducted in accordance with GOST 12088-77. The methodology corresponds to ASTM D737 Standard Test Method for Air Permeability of Textile Fabrics. Measurements were calculated under the following conditions: rarefaction under dotty sample 49 Pa and jaw level for dotty sample 147 N. Air permeability Q (dm^3/s·m^2) was calculated using the equation:

$$Q = \frac{V}{S}$$

where V is average volume of air, dm^3/s; S is measurement area, m^2.

The resistance of the composite shape to wear (storage, loading, axial deformation, dry cleaning) was measured on hemispherical samples with a 10 cm radius formed on a laboratory press from fused fabrics in recommended WHT modes. The samples were stored in a free state at room temperature. They were exposed to static loading for 10 min with a 30 g load (Figure 1), followed by a 5 min rest.

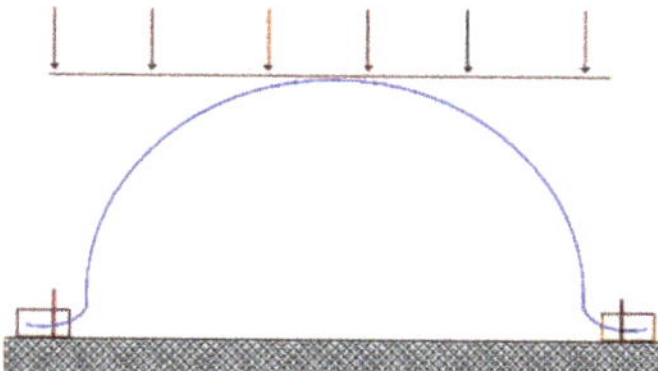

Figure 1. Loading scheme for bulk samples.

The deformation of the hemispherical samples was achieved by stretching them in two perpendicular radial directions at the same rate using a specially developed device that made it possible to mechanize the deformation in this mode creating the necessary conditions for objective studies of the tensile strain. The number of deformation cycles was 20,000. After the deformation, the samples were released from the apparatus clamps and were left in a state of rest for 10 min.

The shape stability factor (SS, %) was calculated by the equation:

$$SS = \frac{H_i}{H_0} \times 100\%$$

where H_0 and H_i are the sample height values 10 min after the sample was formed and after the exposure.

3. Results

3.1. Justification of the Conditions of Two-Component Polymer Coating Formation

Modified polymer coatings can be obtained for a lot of different FIM with a wide variety of thermoplastic compounds (polyolefins, aliphatic and aromatic polyamides, polyethers and polyesters, polyvinyl chloride, polyurethanes, polyvinyl acetate, ethylene, and vinyl acetate copolymers or acrylic compounds) as the thermoplastic adhesive polymer coating (TP). The modifying compound was selected based on its ability to interact with the TP within the range of temperatures that are used in the fusing processes of sewing production, and to ensure that the TP did its job properly. The best results for the most common in garment production polyamide TP were obtained using polyacrylates as the MM. Figure 2 presents the results of controlled interaction of the polyamide adhesive PA-12AKR and oligoacrylate dispersion Akremos 120D as an example.

Using the IR spectroscopy study results (Figure 2a), it was possible to control the type of interactions taking place during the formation of the polyamide–acrylate copolymer under heat treatment. The narrowing of the absorption band (at 725 cm^{-1} in the adduct spectrogram) of the N–H bond stretching vibrations in the polyamide compound amino group was accompanied by the disappearance of the peak at 942 cm^{-1}, corresponding to the double bond stretching vibrations in the CH2=C oligoacrylate molecule. At the same time, the high-intensity absorption on the stretching vibration bands in the C=O and

N–H polyamide chain groups remained the same—860 and 1730 cm^{-1}, respectively, which ensured fabric layer fusion.

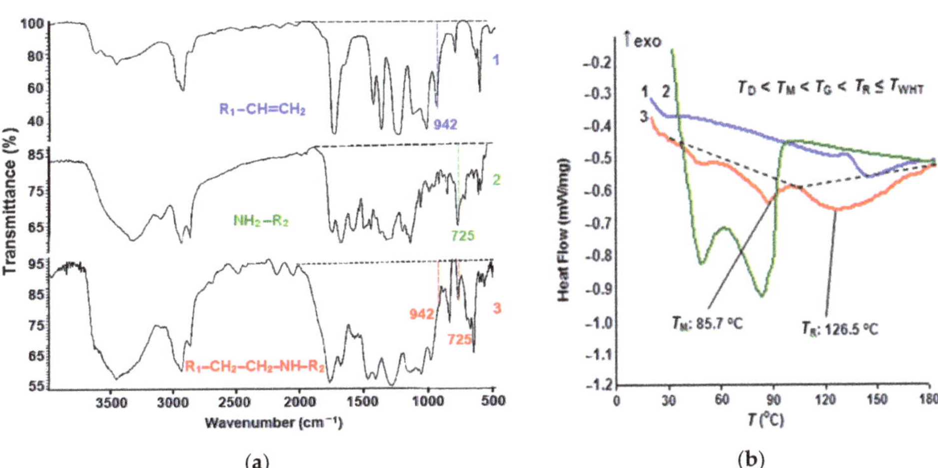

Figure 2. IR spectra (**a**) and DSC curves (**b**) of Acremos 120D oligoacrylate dispersion (1), PA-12AKR polyamide adhesive (2), and polyamide-polyacrylate adduct (3).

The DSC data shown in Figure 2b made it possible to determine the correspondence of the copolymerization temperature peak T_R = 126.5 °C with the WHT heating parameters. To create the optimal manufacturing conditions, the preliminary fusion of the MM and FIM on a thermal press must be carried out at room temperature exceeding the TP melting point in the MP presence (T_M = 85.7 °C), but by no more than 25 °C [17] to prevent considerable spreading of the adhesive dots and, thus, the rigidity increase in the fused fabric.

By analyzing the data in Figure 2, we determined the most suitable modes of the main composite preparation stages. It is proposed to deposit an MP dispersion onto the FIM side coated with an adhesive layer by the screen-printing method, dry it at a temperature below the TP melting point, and then obtain a modified fusible interfacing material (MIM). The following parameters are required to create a fusible coating based on the polyamide PA-12 AKR: MIM drying temperature T_D = 60 °C, MM and MIM fusing temperature T_G = 110 °C, and temperature of the finished garment WHT T_{WHT} = 140 °C.

Table 4 presents measurement results of the bond strength of FIM + MM or MIM + MM fused fabrics obtained at T_G = 110 °C and subjected to WHT at T_{WHT} = 140 °C. The data in Table 4 indicate that the copolymer retained its adhesion properties. The bond strength of MM + MIM fused fabrics was 15%–65% higher than the respective value in the initial standard FIM. The cohesive bond breakage became adhesive, with fibers pulled out of the textile base structure, which indicates that the fibrous materials actively participated in the formation of the highly developed composite interface.

In contrast to the traditional method of formation of a 2D-structured adhesive interlayer that penetrates to a small depth in the interfiber spaces on the surface of the composite layers, graft-copolymers must be introduced into the capillary and pore system of individual fibers. In order to achieve that, a graft polymer dispersion must penetrate into the interfiber pore spaces of the interlining fabric base. It should be borne in mind that swelling of cellulose fibers increases the lateral size of mesoporous cavities to 25–35 nm, with the submicroscopic pore diameter reaching 3–7 nm.

The ability of industrially produced polymer dispersions to penetrate into the structure of the FIM textile base was evaluated using the results of a study of hydrosol particle sizes with the dynamic light scattering method. One of the few compounds satisfying the size requirements was Akratam AS 01, an acrylate dispersion preparation. The characteristics of its colloidal state are shown in Figure 3.

Table 4. The bond strength of fused fabrics.

Fused Fabrics		Bond Strength, P (N/10 cm)		ΔP (%)
MM Type	**FIM**	**MM + FIM**	**MM + MIM**	
MM1	FIM3	4.7	7.1	51
	FIM2	4.1	5.8	41.5
MM2	FIM1	3.7	5.4	46
	FIM4	5.5	7.7	40
MM3	FIM3	4.5	6.6	46.7
	FIM4	5.3	7.8	47.2
MM4	FIM2	4.0	5.6	40
	FIM1	3.6	5.1	41.7
MM5	FIM2	4.3	6.4	48.8
	FIM4	5.4	7.7	42.6

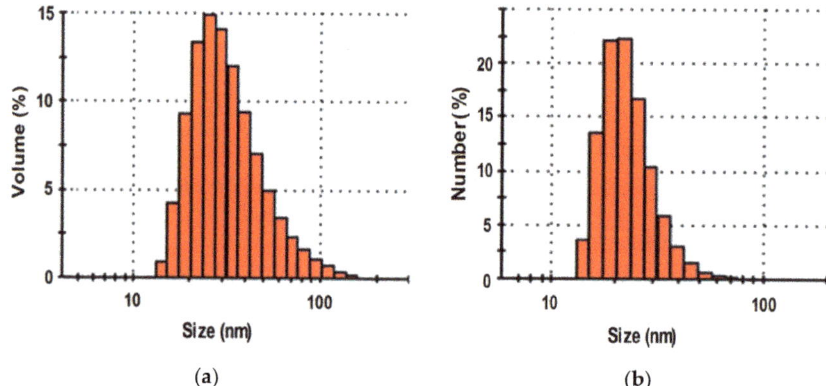

Figure 3. Particle size distribution of the relative dispersed phase volume indicators (**a**) and the particles' relative number (**b**) in the preparation Akratam AS 01 hydrosol.

It is shown that more than 86% of the relative particle number and about 70% of the relative dispersed phase volume were made up by fractions with particles less than 35 nm in size. It means that the compound could migrate into the structure of the cellulose base when the fibrous material gets swollen.

Table 5 shows the results of the particle size analysis of all studied acrylate compounds.

In the vast majority of the studied objects, the size of the particles (r^{MF}) of the major fractions, identified by the total value of the dispersed phase relative volume (V^{MF}) or relative number of particles (N^{MF}), was much bigger than those of the voids in the cellulose fiber capillary pore system. When such compounds are used in their initial form, the dispersed phase particles form agglomerates in the interfiber and interthread spaces of the FIM textile base, which lowers the effectiveness of an adhesive polymer coating modification. The advantages of using compounds with a large share of small-sized fractions consist in the possibility of formation of elongated graft chains in the fibrous material capillaries and the increase in the density of grafting of the side branches to the polymer backbone.

It has been established that it is possible to effectively increase the dispersion degree of oligoacrylate compounds by subjecting them to a series of mechanoacoustic effects: high shear stresses, ultrasound, and cavitation. For example, Figure 4 illustrates changes in the Akremos 120D particle sizes after their processing on a rotary-pulse activator (RPA).

Table 5. Estimation results of the major fractions particlesize (r^{MF}) and relative share (V^{MF}) in hydrosols of MP dispersions.

Type of MP Dispersion	$V = f(r)$ Curve Data		$N = f(r)$ Curve Data	
	r^{MF} (nm)	V^{MF} (%)	r^{MF} (nm)	N^{MF} (%)
Akratam AS 01	21.9–45.6	78.4	18.9–34.0	86.7
Akratam AS 01-M	25.4–45.6	41.5	21.9–34.0	78.3
	171–413	26.3	-	-
Akratam AS 02	99.1–171	82.5	70.9–128	85.4
Acremos 120Д	32.9–110	82.5	31.6–82.1	87.4
Acremos 304	82.1–148	84.0	70.9–128	86.9
Acremos 306	82.1–48	81.8	70.9–128	83.0
Acremos 402	52.9–110	73.5	45.6–82.1	85.6
Anzal KS	198–266	96.1	198–266	94.4

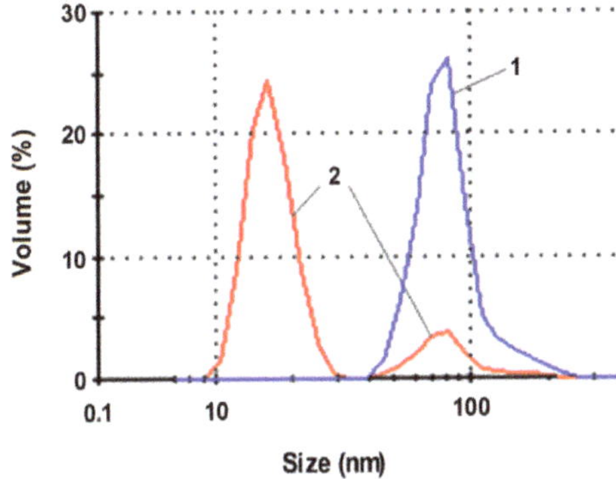

Figure 4. Particle size distribution of the dispersed phase relative volume in the initial dispersion of Acremos 120D (1) and after mechanoacoustic treatment on the RPA apparatus (2).

Hydrosol mechanical activation shifts the peak related to the particle hydrodynamic radius, rmax, from 75 to 15 nm. More than 80% of the dispersed phase relative volume is taken up by the fractions with particles less than 35 nm in size that can be absorbed by the mesoporous cavities of the cellulose fiber. Mechanical activation of the Akratam AS 01 aqueous dispersion produces an ultradispersed hydrosol form with r_{max} = 2.5 nm.

The MP particle size effect on the properties of the fused fabric MM3 + MIM2 composite is illustrated in Figure 5. The rigidity of the composite was compared with that of the fused MM3 + FIM2 composite. Deposition of an oligoacrylate dispersion in its initial form (curve 1) led to a two-fold increase in the EI_K value at the MP content on the material of 0.3 wt.%. Ultrasound treatment aimed at dispersion disaggregation reduced the size of the MP particles to 40 nm (curve 2), which, however, did not make them small enough to penetrate inside the fiber. In this case, 3D-copolymer structures were also formed in the interthread and interfiber spaces of the textile base. Additionally, the EI_K gain achieved by grafting smaller radicals was 17% lower than that reached by deposition of the same amount of the MP in its initial form (see Figure 5).

When MP particles were subjected to mechanically activated grinding and were reduced in size to the mesoporous spaces of the swollen cellulose fiber (curve 3), the rigidity gain (ΔEI_K) relative to that in the initial FIM (G_{MP} = 0) was 3–3.5-fold in comparison to that achieved by deposition of the same amount of the nonactivated dispersion. When ultradispersed MP particles were used (curve 4), they occupied the whole inner volume of the fiber, including the submicroscopic pores, which led to an up to 10-fold increase in ΔEI_K.

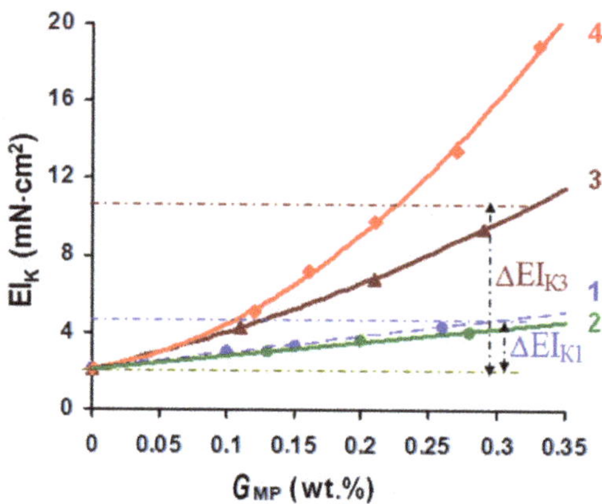

Figure 5. MM3 + MIM2 composite stiffness index (EI_K) dependence on the applied modifying polymer dispersion amount (G_{MP}) with the value of the hydrodynamic particle size r_{max}: 1–75 nm; 2–40 nm; 3–15 nm; 4–2.5 nm.

The data in Figure 5 allow us to describe the concentration dependencies of the rigidity gain with the following expressions (the value in brackets shows the MP dispersion particle size):

$$\Delta EI_{(75nm)} = 8.38 \times G_{MP};$$
$$\Delta EI_{(15nm)} = 18.88 \times G_{MP} + 7.36 \times G_{MP}^2;$$
$$\Delta EI_{(2.5\ nm)} = 11.39 \times G_{MP} + 117.83 \times G_{MP}^2 \tag{1}$$

The dependencies presented reflect the accelerating increase in the rigidity due to the effect of the FIM textile base pore system. Thus, it was reasonable to apply a combination of mechanoacoustic effects to reduce the size of the oligoacrylate particles to 2.5–30 nm to ensure their penetration into the FIM pores.

In order to achieve similar effects on synthetic fabrics (polyester and polyether in particular), which are chemically inert, have a small number of active groups on the surface, are smooth, and have no intrafiber voids, we suggested conducting surface saponification of polyethylene terephthalate in the presence of an interphase catalyst ensuring hydrolysis localization and formation of numerous nano- and microcavities. The role of the catalyst can be played by some quaternary ammonium compounds, in particular Alkamon OS-3.

Figure 6 shows the effects of the saponification conditions on changes in the polyester fiber-free volume (V_P) and stiffness (EI_K) of the composite after fusing and WHT.

As Figure 6 shows, changing the conditions of the polyester fiber saponification stage in the textile base and the amount of the Akratam AS 01 dispersion deposited on the substrate was an effective method of composite rigidity increase control. In comparable experimental conditions, when the compound was used in its initial form and could fill mesoporous cavities, the stiffness gain (ΔEI_K) caused by depositing equal MP amounts could range from 2.3 to 10 times. In the case of the ultradispersed MP form that can penetrate the cavities of submicroscopic sizes, the ΔEI_K value could increase 12.7–16.4 times. The best type of fiber preactivation for treating the initial form of the Akratam AS 01 polymer dispersion was mode 3 ($C_{NaOH} = 1$ mol/L, $\tau = 3$ min). In case of the ultradispersed MP form, surface hydrolysis could be carried out at an alkali concentration lowered to 0.25 mol/L.

3.2. Evaluation of the Polymer Coating Effect on the Processing and Consumer Characteristics of the Obtained Composite

An evaluation of the processing and consumer properties of the composite materials for garment shaping pieces after WHT (finished composite) and semifinished composites at the intermediate stage of fusing (fused fabric) was done. Some of the results for the samples obtained using standard FIM*i* and their modified forms (MIM*i*) are shown in Tables 6 and 7. The MIM samples were prepared on a relatively small area that was coated with the ultradispersed preparation Akremos 120D (S_A = 0.45). The weft threads of the main material and the interlining were aligned with each other during the fusing.

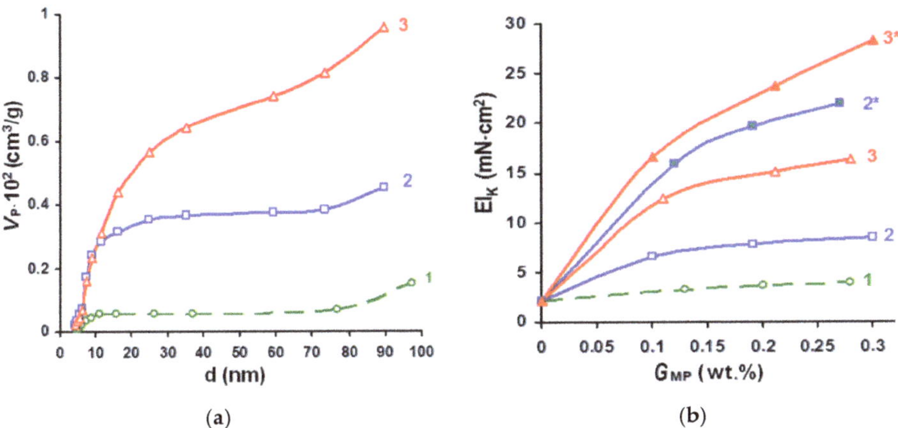

Figure 6. Influence of TPM5 saponification conditions on the dependence of the free volume distribution over the pore diameter (**a**) and changes in the composites stiffness (**b**) with varying the amount of MP applied in the initial (1–3) and ultradispersed (2*–3*) forms: 1-standard TPM5; saponification conditions: 2, 2*-C_{NaOH} = 0.25 mol/L, τ = 10 min; 3, 3*-C_{NaOH} = 1 mol/L, τ = 3 min.

Table 6. Processing properties comparative characteristics (stiffness EI, forming ability *A* and elasticity *U*) of fused fabric (index "FF") and finished composite (index "K") obtained using FIM and MIM.

Composite		EI$_K$ (10^{-3} N·cm^2)		EI$_{FF}$ (10^{-3} N·cm^2)		A_{FF} (%)		U_{FF} (%)		U_K (%)	
		Warp	Weft	Warp	Weft	Warp	Weft	Warp	Weft	Warp	Weft
MM1+	FIM1	12.1	13.4	8.2	9.4	27.1	24.7	39.5	46.4	58.6	68.0
	MIM1	23.9	25.5	8.3	9.7	27.2	24.4	39.7	46.2	79.5	88.2
MM1+	FIM3	9.4	17.1	6.6	11.5	28.5	27.4	39.4	48.5	56.2	72.1
	MIM3	21.2	28.5	6.3	11.1	28.7	27.5	39.1	48.6	77.2	91.6
MM1+	FIM4	14.8	23.4	9.7	14.3	26.6	21.3	42.0	51.2	61.8	74.7
	MIM4	26.1	35.5	9.5	14.5	26.3	21.5	41.8	51.0	80.3	94.2
MM4+	FIM1	9.1	9.4	6.1	6.5	32.6	29.5	35.5	46.0	50.0	57.1
	MIM1	21.2	21.4	5.8	6.3	32.8	29.5	35.1	45.7	72.5	78.6
MM4+	FIM3	7.6	12.6	5.1	8.5	32.9	31.0	34.3	39.1	49.0	55.8
	MIM3	19.4	23.8	4.9	8.6	32.7	31.2	34.2	39.1	71.5	77.3
MM4+	FIM4	10.8	16.7	7.1	11.1	30.1	27.8	37.3	42.8	53.9	62.0
	MIM4	22.2	27.8	7.3	10.7	30.3	27.6	37.2	43.1	75.4	82.5

An analysis of the initial parameters of the objects to be modified considering the processing characteristics of the FIM, given in Table 2, was done. In all of the suite fabrics, we observed the same types of regularities, namely an increase in the stiffness and elasticity of the composite along the weft aligned with the reinforcing weft yarn in the FIM structure. The EI$_K$ and U_K values also became higher when the mass of the textile base and the M_S and S_{TP} indicators of thermoplastic adhesive were increased. The 1.9-fold growth in the adhesive coating area following the replacement of FIM1 with FIM4 led to a 1.75-fold stiffness increase along the weft despite a 7% decrease in M_S value. Along the warp (when FIM did not have a reinforcing thread), the increase in stiffness along the warp value following the S_{TP} growth was smaller (1.22-fold) and more sensitive to changes in the interlining fabric mass.

Table 7. Consumer properties of composites based on FIM and MIM: thickness h; shape stability factor for storage in 24 h (SS_{24}), 48 h (SS_{48}), loading (SS_{load}), multiaxis cyclic tensile strain ($SS_{tensile}$), and dry cleaning ($SS_{dry\ clean}$); air permeability Q and hygroscopicity Hyg.

Composite		h (mm)	Shape Stability Factor (%)					Q (dm³/s m²)	Hyg (%)
			SS_{24}	SS_{48}	SS_{load}	$SS_{tensile}$	$SS_{dry\ clean}$		
MM1+	FIM1	1.90	85	84	75	70	74	84.3	5.3
	MIM1	1.86	90	89	81	77	80	83.9	5.4
MM1+	FIM3	1.76	87	85	76	72	76	78.1	5.5
	MIM3	1.75	93	91	82	79	81	77.8	5.7
MM1+	FIM4	1.76	91	88	82	80	85	50.1	5.1
	MIM4	1.75	96	94	88	83	91	50.1	5.1
MM4+	FIM1	2.0	77	73	64	61	63	108.5	4.4
	MIM1	1.95	81	78	74	69	68	107.9	4.6
MM4+	FIM3	1.90	76	73	59	60	63	84.3	4.5
	MIM3	1.85	80	78	73	67	69	84.4	4.7
MM4+	FIM4	1.80	80	78	69	65	70	52.1	4.2
	MIM4	1.80	85	84	77	72	75	52.5	4.3

When standard FIM were used, the maximum EI_K value was 23.4×10^{-3} N·cm². To reach this value, the MM1 garment pieces had to be oriented crosswise, along the reinforcing weft thread of FIM4, the processing characteristics of which were close to the extremely high values: $M_S = 75$ g/m²; $G_{WT} = 60.8\%$; $S_{TP} = 25.2\%$. A further increase in the rigidity in order to fix the garment shape could be only achieved by using several FIM layers.

The main purpose of FIM polymer coating modification is increasing the EI_K level in order to reduce the number of layers in the fused composite and lower materials' consumption. A comparison of the data in Tables 1 and 6 showed that the EI_K gain relative to the main material rigidity (EI_{MM}) increased from 2.2–7.6 times in the fabrics with standard FIM to 5.1–11.5 times in the used MIM variants.

After the fusion stage, the rigidity of the standard fused fabric composites (EI_{FF}) became 1.5–4.8 times higher than that of EI_{MM}. The higher rigidity of semifinished product had a negative effect on their ability to take the required shape. In the samples with FIM, the decrease in the fused fabric shaping ability A_{FF} in comparison with the A_{MM} reached approximately 35.6%. In spite of the good structural mobility of knitted FIM, the use of molten adhesive dots to fix their parts on the MM surface made the fused composite elasticity quite high. The U_{FF} value along the reinforcing weft threads of the FIM exceeded 50%. Since excessive elasticity of fabrics can be an obstacle to garment shaping from flat pieces, it is desirable to prevent its growth in a fused fabric.

This means that an important component of interlining polymer coating functionalization is ensuring the timeliness of copolymerization in order to preserve the maximum plasticity of the fused composite and obtain the necessary 3D shapes with their subsequent fixation at the final WHT stage. The data in Table 6 provides clear evidence that the use of finely dispersed graft forms and their introduction into the pore structure of the MIM textile base fibers have no effect on the stress-strain properties of the fused fabric. The EI_K/EI_{FF} ratio value rose from 1.4–1.6 in FIM-based fused fabrics to 2.5–4 in the MIM-based ones.

An equally important aim of the interlining polymer coating functionalization is higher elasticity of finished composite (U_K) for shape stability under external effects. A comparison of the values of the EI_K^{weft} and U_K^{weft} indicators for FIM1 and FIM3 fabrics showed that the higher rigidity of the standard interlining materials did not always make their elasticity higher as a result of the action of several structural factors. The proposed technology of forming a highly branched composite interface with the side chains of the graft-copolymer penetrating into the intrafiber pore spaces of the fusing textile layers allowed a simultaneous increase in EI_K and U_K indicators. The relative elasticity gain after the WHT (U_K/U_{FF}) rose from 1.2–1.5 times in the fused fabrics with FIM to 1.7–2.1 times in the samples with MIM.

Dependencies have been obtained showing how the processing parameters of the main material (the MM index) and FIM structural parameters—surface density (M_S, g/m^2), weft thread mass fraction (G_{WT}, %), and adhesive coating area (S_{TP}, %)—affect the changes in the respective properties of the composite obtained using standard FIM:

$$
\begin{cases}
EI_K^{warp} = EI_{MM}^{warp} \times (0.029 \times M_S + 0.049 \times S_{TP}); \ R^2 = 0.979; \\
EI_K^{weft} = EI_{MM}^{weft} \times (0.005 \times M_S + 0.002 \times G_{WT} + 0.27 \times S_{TP}); \ R^2 = 0.982; \\
EI_{FF} = (0.67 \dots 0.72) \times EI_K; \\
A_{FF}^{warp} = \dfrac{A_{MM}^{warp}}{(0.012 \times M_S + 0.017 \times S_{TP})}; \ R^2 = 0.987; \\
A_{FF}^{weft} = \dfrac{A_{MM}^{weft}}{(0.011 \times M_S + 0.01 \times G_{WT} + 0.323 \times S_{TP})}; \ R^2 = 0.984; \\
U_K^{warp} = 44.5 + 0.14 \times M_S + 0.26 \times S_{TP}; \ R^2 = 0.88; \text{—only for MM1} \\
U_K^{weft} = 45 + 0.0186 \times M_S + 0.095 \times G_{WT} + 0.65 \times S_{TP}; R^2 = 1; \text{—only for MM1} \\
U_{FF} = (0.65 \dots 0.75) \times U_K
\end{cases}
\tag{2}
$$

The data in Table 6 indicate that deposition of a modifying dispersion had practically no effect on the processing characteristics of the obtained fused fabric. Hence, Equation (2) used for EI_{FF}, A_{FF}, and U_{FF} indicators can be also applied to MIM.

Table 7 shows the effect of the interfacing polymer coating modification on changes in the consumer properties of the obtained composite samples.

The composites with MIM were found to be thinner than those with FIM. It was, evidently, the result of the denser fiber structure of the interfacing material caused by MP penetration into the pores and formation of a developed interphase layer. This also made the shape stability factor (SS) of the obtained MIM-based finished composite noticeably higher both in storage and in wear. It should be said that the SS reduction on the second day of storage in both types of interlining materials was rather small, i.e., the shape relaxation took place in the first 24 h after the fabric was formed, and then the shape remained stable. The shape stability values for the composites correlated with the changes in their elasticity. In the FIM-based samples, the correlation between the indicators can be described by the following set of equations:

$$
\begin{aligned}
SS_{24} &= 0.71 \times U_K^{warp} + 0.64 \times U_K^{weft}; \ R^2 = 0.989; \\
SS_{load} &= 0.7 \times U_K^{warp} + 0.5 \times U_K^{weft}; \ R^2 = 0.989; \\
SS_{tensile} &= 0.5 \times U_K^{warp} + 0.62 \times U_K^{weft}; \ R^2 = 0.985
\end{aligned}
\tag{3}
$$

The shape stability factors for the MIM-based composites can be described by the following set of equations:

$$
\begin{aligned}
SS_{24} &= 0.071 \times U_K^{warp} + 0.96 \times U_K^{weft}; \ R^2 = 0.992; \\
SS_{load} &= 0.36 \times U_K^{warp} + 0.61 \times U_K^{weft}; \ R^2 = 0.981; \\
SS_{tensile} &= 0.14 \times U_K^{warp} + 0.76 \times U_K^{weft}; \ R^2 = 0.988
\end{aligned}
\tag{4}
$$

Hygiene safety of composite materials for making clothes is evaluated by breathability and hygroscopicity. Fabric breathability is important for maintaining the optimal temperature in the undergarment space and for diversion of carbon dioxide vapor intensively released by human skin (about 250 mg/h). The Q indicator values for composite samples were 55%–58% lower than the initial level for the MM, as fabric bonding dramatically increases the resistance to the air movement. The increase in FIM surface density and TP mass was accompanied by a considerable reduction in the composite Q value. The macropores in the composite structure were blocked by the adhesive interlayer more quickly as the adhesive coating area S_{TP} became larger. The correlation between the values is described by the equation:

$$Q = \frac{Q_{MM}}{0.019 \times M_S - 0.014 \times S_{TP} + 0.0042 \times S_{TP}^2}; \quad R^2 = 0.987 \tag{5}$$

An important fact is that MP dispersion introduction into the fibrous material structure and formation of a graft-copolymer adhesive form in fused composites with MIM did not lead to additional resistance of the composite garment pieces to air penetration. At the same time, the effectiveness of the polymer coating modification could be evaluated by determining the EI_K/Q indicator ratio, which becames 1.5–2.6 times higher when standard FIM were replaced with developed MIM.

Hyg indicator values are interrelated with the content and hydrophilicity of fibrous components of fused fabric layers. The presence of a TP polymer matrix in the composites lowered the Hyg value by 50%–67% compared to the same value for MM. However, in all the samples compared, the use of MIM had practically no effect on the basic Hyg values corresponding to comfortable wear conditions.

It is especially valuable that MIM makes it possible to regulate the rigidity of various parts of a composite garment-shaping piece without using multilayered laminated fabrics. This result can be achieved by using screens with changing printing patterns corresponding to the required reinforcement area (S_A) for screen printing of the modifying dispersion (see Table 3). In order to be able to regulate the rigidity, we derived mathematical dependencies describing changes in the main processing parameters of the composite depending on the S_A value (with the first member of the equation right side reflecting the indicator value in the fused composites with a nonmodified interfacing fabric):

$$\begin{aligned} EI_K^{MIM} &= EI_K^{FIM} - 7.73 \times S_A + 76.84 \times S_A^2 \; (\text{range } \Delta EI_K = (6.7\text{–}27.4) \times 10^{-3} \, \text{N·cm}^2); \\ U_K^{MIM} &= U_K^{FIM} + 39.85 \times S_A + 19.72 \times S_A^2 \; (\text{range } \Delta U_K = 16.4\%\text{–}34.2\%). \end{aligned} \tag{6}$$

An analysis of the equations shows that by regulating S_A, it is possible to change the composite rigidity (ΔEI_K) and elasticity (ΔU_K) within a wide range of values.

3.3. Evaluation of Effects of Nanodispersed Fillers

One of the reinforcing nanomodifier types widely used in composite material production is detonation nanodiamonds (DND). Figure 7 shows the results of a comparison of the stress-strain properties of an MM1-based composite with those of standard FIM1 and its modified form with different amounts of DND introduced into the Akremos 120D oligoacrylate dispersion. The mechanically activated binary composition was deposited onto the interfacing material using screen No. 1 ($S_A = 0.35$).

Within the studied range of DND concentrations, the stiffness gain in the modified composite relative to the initial fused composite (ΔEI_{DND}) was 1.7–2.6 times bigger than in the case of using oligoacrylate without a filler (ΔEI_0), and the stiffness gain ΔU_K was 1.24–2.2 times bigger. The changes were more noticeable at the C_{DND} up to 5 wt.%, when the ΔEI_K value doubled and ΔU_K increased 1.75-fold. A further increase in the DND concentration slowed down the indicator growth. This was probably caused by the negative DND particle surface zeta-potential. The higher modifier concentration increased the total charge value of the polymer dispersion and strengthened the forces of electrostatic repulsion with the negatively charged surface of the viscose fiber in the interlining.

The data in Figure 8 illustrate the limitations in the use of DND for modifying the interfacing polymer coating. The DND presence triggered agglomeration of the polymer dispersion particles, which, after 24-h storage, produced fractions 100–400 nm in size and micrometer fractions that took up more than 30% of the dispersed phase volume.

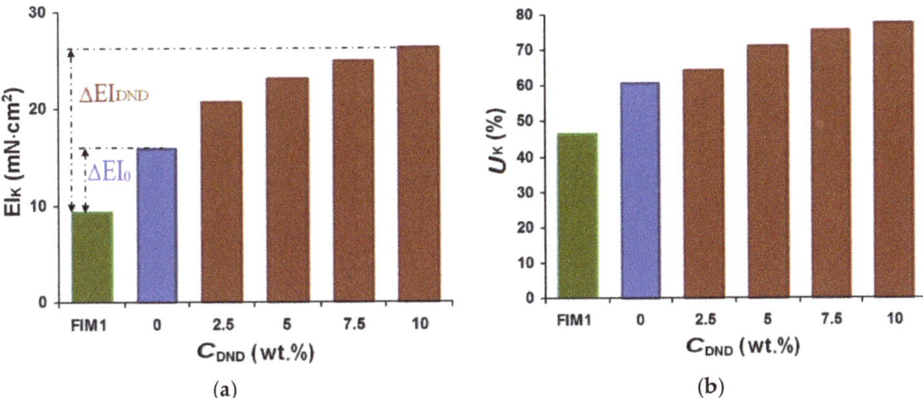

Figure 7. Dependencies of the composite stiffness (**a**) and elasticity (**b**) on the DND addition to the oligoacrylate dispersion (C_{DND}, wt.%).

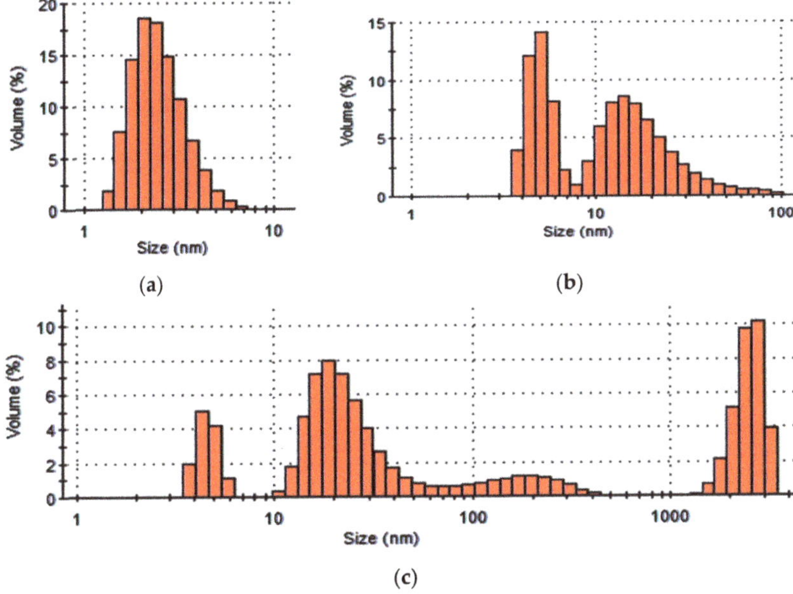

Figure 8. Particle size distribution of the dispersed phase relative volume in the DND preparation hydrosol (**a**) and its mixture with the Acremos 120D preparation after 1 h (**b**) and 24 h (**c**).

Since the polymer dispersion hydrosols are stable in colloidal state and the market prices of nanomodifiers are quite high, it is thought to be more effective to use available nanodispersed silicon dioxide (SD) compounds.

The data in Figure 9 illustrate the necessity of preliminary dispersion of colloidal SD compounds as the initial hydrosol form contained large aggregated fractions of 60–180 nm in size.

Ultrasound effects ensure hydrosol dispersion, with most of the volume occupied by fractions of individual SD grains capable of penetrating into the mesoporous spaces of the interlining fabric fibrous base. The SD ultradispersion effect was achieved by applying a combination of ultrasound, high-rate shear loads, and cavitation in the RPA apparatus. About 40% of the relative dispersed phase volume was taken up by the particles of less than 10 nm in size. Evidently, mechanoacoustic effects led to crushing of the SD grains, which is illustrated by the micrographs with damaged spherical parts in Figure 10.

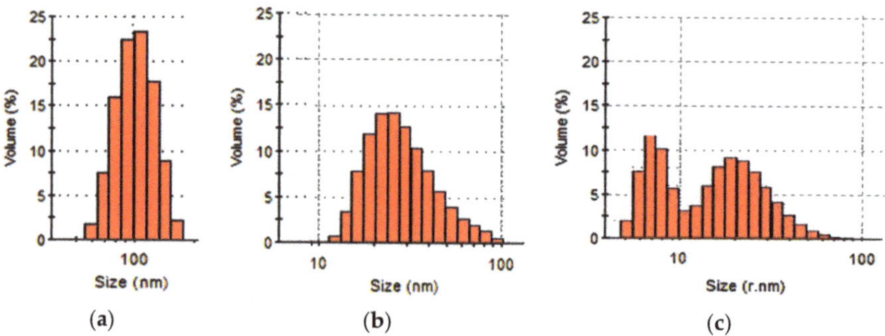

(a)   (b)   (c)

Figure 9. Changes in the fractional distribution of the initial SD hydrosol particles (**a**) after ultrasonic treatment (**b**) and mechanical activation in RPA (**c**).

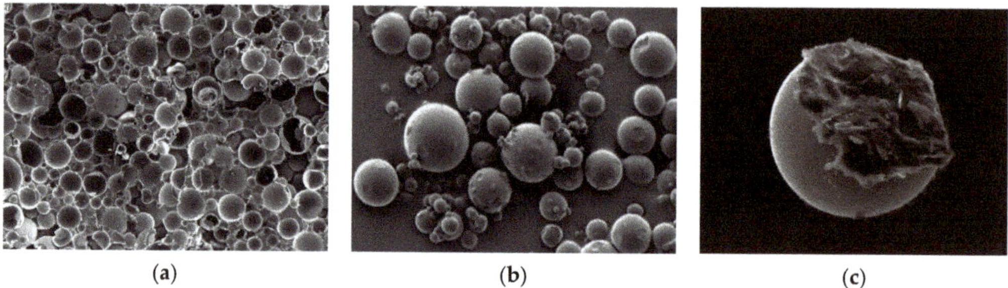

(a)   (b)   (c)

Figure 10. SEM images (**a–c**) of colloidal SD after mechanoacoustic processing in RPA.

The hypothesis about SD grain crushing caused by mechanical activation of the hydrosols and SD's ability to interact with oligoacrylates of different structures is confirmed by IR-spectroscopy studies [21]. The specifics of the interactions in the system are illustrated by the data in Figure 11.

The absorption bands for the SD compound reflected in curve 1 are reproduced in curve 2, which indicates that the chemical state of the colloidal SiO_2 remained unchanged after the US-treatment. Curve 2 also contained a group of oligoacrylate molecule bands: the stretching vibrations of the C=O groups in the butylacrylate fragments at 860 and 1692 cm^{-1}, the C–C stretching and skeletal vibrations at 1512 and 1600 cm^{-1}, as well as the bending (910 cm^{-1}) and stretching (2850 cm^{-1}) vibrations of the C–H bond in the alkyl chain. The high-intensity band appearing at 2156 cm^{-1} on curve 2 was formed by the vibrations of the silanol groups participating in the formation of a hydrogen bond with the acrylate carbonyl. This indicates the physical nature of the adsorption interactions between the system components subjected to US-dispersion.

When the components were treated together in the RPA, it caused a considerable transformation of the spectrum (curve 3). Changes in the SD state were reflected in the lower intensity of the Si–O stretching vibration bands at 465, 630, 800, 1194 and 1960 cm^{-1} and bending vibrations (at 486 and 560 cm^{-1}) in the O–Si–O bridge bonds, which indicated mechanically activated breakage of the siloxane bonds in the SiO_4 tetrahedron network. The higher intensity of the Si–OH bending vibrations peak at 870 cm^{-1} indicated an increase in the number of silanol groups.

Polymerization triggered by mechanical activation of the binary system led to the disappearance of the vinyl group rocking vibration band on curve 3 (730 cm^{-1}) and peaks of the CH_2=C– double bond stretching vibrations (3044, 1721, 1550, 942 cm^{-1}) on curve 2. The band at 3044 cm^{-1} on curve 3 shifted to the higher frequency region (3082 cm^{-1}) as a result of the increasing absorption of the stretching vibrations in the CH_2 and CH_3 groups formed by the oligoacrylate vinylidene unit transformation.

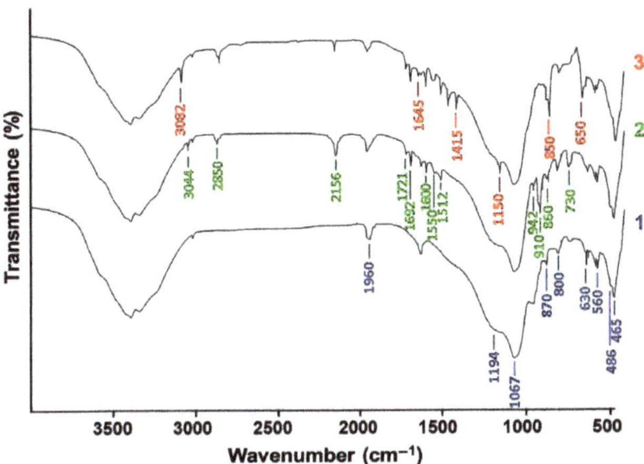

Figure 11. IR spectra of SD (1) and SD-Akratam AS compositions in a ratio of 7:3, obtained using ultrasound processing (2) and mechanical activation (3) of a binary hydrosol.

However, curve 3 had bands indicating the formation of new bond types. The band at 1150 cm^{-1} demonstrated the appearance of a siloxane group bond with a Si–O–C carbon atom. The peaks at 650 and 850 cm^{-1} indicated the formation of a Si–C bond. There were also absorption bands at 1415 and 1645 cm^{-1} attributed to the scissoring vibrations of the hydrogen atoms in the Si–(R)C<(H)$_2$ groups. The results of the IR-spectra analysis allow us to conclude that subjecting the preparations simultaneously to the action of shear stress, ultrasound, and cavitation led to a break-up of the siloxane bonds in the structure of the silica dioxide nanoparticles and was accompanied by broken bond hydrolysis. At the same time, it is quite likely that Si–O–Si bond breakage may result in the formation of radical products, and the siloxane macroradical may get attached to the oligoacrylate "tail", which can facilitate stabilization of the binary hydrosol state and more uniform SD distribution within the polymer adhesive bulk.

Figure 12a presents the results of the thermal analysis of the two-component systems modelling copolymerization of oligoacrylate with a standard FIM polyamide adhesive.

The peak at 76.2 °C on curve 1 characterized the melting of the polyamide not interacting with the acrylate. Film formation started at 118.3 °C, whereas copolymerization in the presence of an initiator-halogenide complexonate of transition metals-and hardening of the adhesive completed at 186.7 °C. In the acrylate-SD composition (curve 2), thermally fusible polymer melting was contained, which made it possible to raise the drying temperature of the FIM being modified (see Figure 2) to T_D = 85–90 °C. At the fusion stage, the heating must be carried out at temperatures not exceeding T_G = 105–115 °C. Polyamide film and graft-copolymer formation processes represented one stage with a peak at 129 °C, which determines the minimum value of the T_{WHT} parameter for the final WHT in order to shape the garment and stabilize its shape.

Figure 12b demonstrates the increase in the rigidity of the obtained composite as the SiO_2 content in the acrylate dispersion became higher. By changing the nanodispersed filler content from 5 to 25%, it was possible to increase the rigidity in composite garment shaping pieces 1.08–1.8 times, which may be a new solution to the problem of quick adjustment of the properties of interlining materials to the requirements for garment designs being developed.

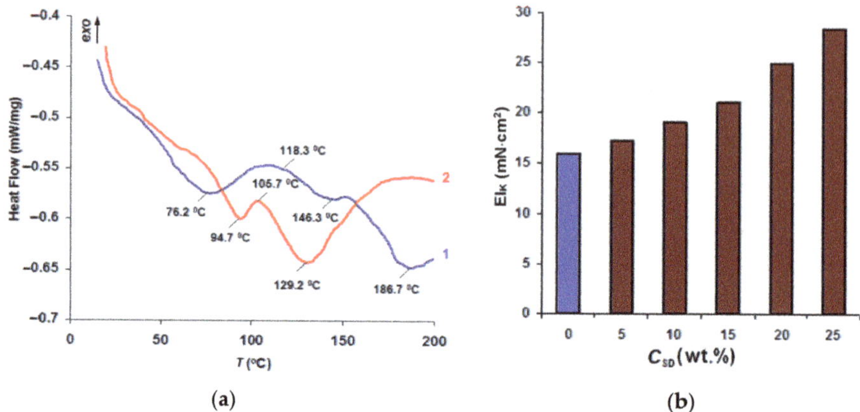

Figure 12. DSC analysis curves of PA-12AKR co-crystallization adducts of glue with the initial Akratam AS (1) and the acrylate-SD composition in a ratio of 3:1, (2) (**a**) and the stiffness index of the finished composite dependence on the SD concentration (**b**).

4. Discussion

The presented results indicate that a modified polymer coating ensured the formation of a graft-copolymer interface in the composite structure, which can have a great effect on the stress-strain properties of the garment piece made. The main difference of the developed MIM from traditional FIM was the replacement of the method consisting in forming 2D-structured adhesive layers between the fabrics to be joined to creating composites with a highly branched 3D interface. As Table 6 and Equations (2) and (5) show, the main technology used to make the rigidity of one type of the FIM textile base higher was increasing the adhesive dot area (TP mass), which was accompanied by a considerable reduction in the breathability of the finished composite. When MIM with a modified polymer coating were used, both the rigidity and the elasticity grew, which quite naturally led to higher strength and durability (see Equation (4)).

To demonstrate the effectiveness of using a modified coating, we calculated the main parameters of the fused fabric and finished composite obtained from MM1 fabric with FIM1 or MIM1 in garment pieces with the same preset value of the EI_K indicator along the MM warp yarn. In case of FIM, we used Equation (2) to determine the necessary adhesive coating area S_{TP}. For MIM, we used Equation (5) to select the MP deposition area S_A. The results of the comparison of the calculated values for the fused fabric and finished composite are given in Table 8.

Table 8. Comparison of the fused fabrics and composites based on FIM and MIM characteristics calculated values.

Required Value EI_K (10^{-3} N·cm²)	Type of Interlining Material	Conditions for Creating EI_K		A_{FF} (%)	U_K (%)	Q (dm³/s m²)
		S_{TP} (%)	S_A			
12	FIM1	13.2	-	30.7	58.6	84.3
17	FIM1	35.8	-	19.3	65.4	28.6
	MIM1	13.2	0.31	30.7	72.8	84.3
22	FIM1	60.3	-	15.3	72.8	11.5
	MIM1	13.2	0.41	30.7	78.3	84.3
27	FIM1	84.8	-	12.6	80.1	6.0
	MIM1	13.2	0.5	30.7	83.5	84.3
32	MIM1	13.2	0.56	30.7	87.1	84.3
40	MIM1(C_{SD} 7.1%)	13.2	0.31	30.7	86.7	84.3

It was found that the range of possible rigidity changes in the fabric with FIM1 as a result of increasing the TP amount was extremely limited. The increase of 5×10^{-3} N·cm^2 in EI$_K$ could only be achieved if the coating area, S_{TP}, made up 35.8%; and the value growth of 10×10^{-3} N·cm^2 could be only reached if the area occupied by the coating was 60.3%, which critically reduces the shaping ability and breathability. It should be said that the A_{FF} reduction by less than 18% and the U_K increase by more than 50% worsened the shaping conditions at the WHT stage and could cause a greater number of process defects in the form of creases and wrinkles on the shell fabric surface and quick shape relaxation. This means that it is possible to prepare composites with rigidity of $(20–40) \times 10^{-3}$ N·cm^2 by the traditional method only if several FIM layers are used.

The use of MIM made it possible to achieve the required value of the EI$_K$ indicator by modifying the interlining fabric at a minimum value of S_{TP}, even without reaching extremely high S_A values. MIM-based composites have high values of all the parameters. This allows us to claim that MP deposition facilitates preparation of composites exceeding the traditional types in all the processing characteristics and consumer properties.

As an example, it was also shown that the maximum rigidity value $(39.8 \times 10^{-3}$ N·cm$^2)$ that is required to ensure shape stability of the shoulder part of a large-size men's jacket of rigidly fixed form can be achieved by using an MP with a nanofiller. When the concentration of the additive, C_{SD}, in the composition with Akratam AS reached 7.1%, the target parameters of the composite could be obtained at the minimum values required for polymer coating modification (S_{TP} and S_A).

Thus, the multifunctionality of the interfacing polymer coating was the result of the improvement of the processing characteristics and consumer properties of fused fabric and finished composite. The methods of screen or ink-jet printing with an aqueous MP dispersion made it possible to vary the composite rigidity by changing the S_A value. The increase in the reinforcement area was accompanied by higher elasticity but did not lead to lower bond strength and did not reduce the shaping ability and breathability either. Changing the pattern of MP deposition over the MIM area made it possible to introduce gradient changes in the rigidity value within one reinforced garment piece. Higher reinforcement could be achieved by applying methods of nanostructural organization of MIM polymer coating. Easy-to-use methods of hydrosol ultradispersion and nanolayer surface modification of fibers were applied to ensure effective MP penetration into the intrafiber pore structure of textile bases.

MIM functionality could be widened by using nanodispersed fillers for fabric reinforcement or by providing special-purpose clothes with health improvement properties. An evaluation of the effect of the interface modification by ultradispersed silicon dioxide forms and detonation nanodiamonds showed that it was possible to increase the composite rigidity and elasticity. The higher elasticity (1.05–1.2-fold increase) in this case was more important, as it directly correlated with the resistance of a finished garment shape to wear.

It was found that an oligoacrylate graft dispersion could be used to stabilize the disperse state of nanosized fillers, to make their distribution in the composite structure more uniform and to achieve reproducible technological effects by modifying the interlining polymer coating. The proposed method of simultaneous mechanical activation of a nanodispersed filler with an MP hydrosol could be applied to immobilize highly coercive nanoparticles, strontium ferrite in particular, in a fibrous material. The effectiveness of the developed method in producing magnetic fabrics generating a constant low-intensity magnetic field in the near-surface layer is shown in [13]. Special garment design [14] improves the adaptation and regeneration capacity of healthy people in psychologically or physically stressful situations and makes it possible to apply magnetic therapy for treatment and prevention of many diseases.

5. Conclusions

The presented results indicate that the technology proposed by us for modifying polymer coatings of fusible interlining materials by depositing a grafted oligomer dispersion capable of penetrating into the intrafiber pore structure of the textile base can be used to widen the functionality of interlining materials and composite garment pieces based on them. The nanostructured architecture of the modified polymer coating and timely (at the final WHT stage) oligoacrylate copolymerization with the macromolecules of a thermoplastic adhesive facilitate the formation of highly branched 3D-structured interfaces in the composites, which considerably increases the processing characteristics and consumer properties of the finished garments in comparison with those made with standard FIM.

The studied methods of ultradispersing polymer dispersions and nanolayer fibrous substrate surface modification complement each other and make it possible to control the area of formation of a highly developed adhesive structure and to reduce the amount of reinforcing nanomodifiers to reasonable values. Together with screen printing methods, in which it is possible to change the area of modifying composition deposition, they allow for the development of a wide variety of interlining fabrics from a small range of standard FIM and quick adjustment of their functional characteristics in order to produce clothes of any shape and with any elasticity degree.

6. Patents

Patent RU 2,383,672 C2. Compound for giving stability of shape to apparel components. Publ. 10.03.2010.

Author Contributions: Conceptualization, N.K. and S.K.; methodology, S.K.; validation, A.B., E.N. and N.K.; formal analysis, O.R. and S.A.; writing—original draft preparation, N.K. and S.K.; writing—review and editing, A.B.; visualization, O.R. and S.A.; supervision, A.B.; project administration, E.N. All authors have read and agreed to the published version of the manuscript.

Funding: This research received no external funding.

Institutional Review Board Statement: Not applicable.

Informed Consent Statement: Not applicable.

Data Availability Statement: Data is contained within the article.

Conflicts of Interest: The authors declare no conflict of interest.

References

1. Phebea, K.; Krishnaraj, K.; Chandrasekaran, B. Evaluating performance characteristics of different fusible interlinings. *Indian J. Fibre Text. Res.* **2014**, *39*, 380–385.
2. Zhang, Q.; Kan, C.-W.; Chan, C.-K. Relationship between physical and low-stress mechanical properties to fabric hand of woollen fabric with fusible interlinings. *Fibers Polym.* **2018**, *19*, 230–237. [CrossRef]
3. Zhang, Q.; Kan, C.-W. A review of fusible interlinings usage in garment manufacture. *Polymers* **2018**, *10*, 1230. [CrossRef] [PubMed]
4. Kim, K.; Takatera, M. Effects of dot-type adhesive and yarn float on shear stiffness of laminated fabric with interlining. *Text. Res. J.* **2015**, *86*, 480–492. [CrossRef]
5. Koksharov, S.A.; Kornilova, N.L.; Shammut, Y. Design of composite materials for clothes. In Proceedings of the International Scientific Practical Conference "Materials Science, Shape-Generating Technologies and Equipment 2020" (ICMSSTE 2020), Yalta, Russia, 25–29 May 2020; p. 03001.
6. Amar, Z.; Al-Gamal, G. Effect of different types and orientations of fusible interlinings on men striped shirt cuffs. *J. Am. Sci.* **2015**, *11*, 66–72.
7. Yun, S.Y.; Kim, S.M.; Park, C.K. Development of an expert system for optimum fusible interlining. *Fash. Text. Res. J.* **2009**, *11*, 648–660.
8. Zhang, Q.; Kan, C.-W. Property comparison of woollen fabrics with fusible and printable interlinings. *Fibers Polym.* **2018**, *19*, 987–996. [CrossRef]
9. Filippov, A.P.; Belyaeva, E.V.; Zakharova, N.V.; Sasina, A.S.; Ilgach, D.M.; Meleshko, T.K.; Yakimansky, A.V. Double stimuli-responsive behavior of graft copolymer with polyimide backbone and poly(n,n-dimethylaminoethyl methacrylate) side chains. *Colloid Polym. Sci.* **2015**, *293*, 555–565. [CrossRef]
10. Simonova, M.A.; Zamyshlyayeva, O.G.; Simonova, A.A.; Filippov, A.P. Conformation of the linear-dendritic block copolymers of hyperbranched polyphenylenegermane and linear poly(methylmethacrylate). *Int. J. Polym. Anal. Charact.* **2015**, *20*, 223–230. [CrossRef]
11. Vaziri, H.S.; Omaraei, I.A.; Abadyan, M.; Mortezaei, M.; Yousefi, N. Thermophysical and rheological behavior of polystyrene/silica nanocomposites: Investigation of nanoparticle content. *Mater. Design* **2011**, *32*, 4537. [CrossRef]

12. Kontou, E.; Anthoulis, G. The effect of silica nanoparticles on the thermomechanical properties of polystyrene. *J. Appl. Polym. Sci.* **2007**, *105*, 1723. [CrossRef]
13. Izgorodin, A.K.; Patrusheva, T.N. Magnetic fabric: Development of component composition and production technology. *Russ. J. Gen. Chem.* **2013**, *83*, 169–176. [CrossRef]
14. Izgorodin, A.K. Protective Clothing for Rescue and Salvage Operations. Ru. Patent 2448622 C2, 25 March 2010.
15. Koksharov, S.A.; Kornilova, N.L.; Shammut, J.A.; Radchenko, O.V. Synthesis of a highly chained polymeric connecting in the structure of a multilayered package for garments. *Key Eng. Mater.* **2019**, *816*, 219–227. [CrossRef]
16. Koksharov, S.A.; Kornilova, N.L.; Fedosov, S.V. Development of reinforced composite materials with a nanoporous textile substrate and a brush-structured polymer interfacial layer. *Russ. J. Gen. Chem.* **2017**, *87*, 1428–1438. [CrossRef]
17. Kornilova, N.; Koksharov, S.; Arbuzova, A.; Shukla, A.; Mundkur, S.D. Development of reinforced interlining materials to regulate elastic properties. *Indian J. Fibre Text. Res.* **2017**, *42*, 150–159.
18. Kornilova, N.L.; Koksharov, S.A.; Shammut, U.A.; Radchenko, O.V.; Nikiforova, E.N. Influence of dispersity of reinforcing polymer to the polymerfiber composite materials' rigidity. *J. Phys. Conf. Ser.* **2020**, *1451*, 012012. [CrossRef]
19. Yan, Y.D.; Clarke, J.H.R. In-situ determination of particle size distributions in colloids. *Adv. Colloid Interface Sci.* **1989**, *29*, 277–318. [CrossRef]
20. Koksharov, S.A. On the application of the dynamic light scattering method for estimating the size of nanoparticles in bicomponent hydrosol. *Russ. J. Chem. Chem. Technol.* **2015**, *58*, 33–36.
21. Aleeva, S.V.; Koksharov, S.A.; Kornilova, N.L. Interactions in mechanoactivated hydrosols of colloidal silica and oligoacrylates. *Russ. J. Phys. Chem. A* **2020**, *94*, 1268–1271. [CrossRef]

Article

Properties of Polypropylene Yarns with a Polytetrafluoroethylene Coating Containing Stabilized Magnetite Particles

Natalia Prorokova [1,2,*] and Svetlana Vavilova [1]

[1] G.A. Krestov Institute of Solution Chemistry of the Russian Academy of Sciences, Akademicheskaya St. 1, 153045 Ivanovo, Russia; sjv@isc-ras.ru
[2] Department of Natural Sciences and Technosphere Safety, Ivanovo State Polytechnic University, Sheremetevsky Ave. 21, 153000 Ivanovo, Russia
* Correspondence: npp@isc-ras.ru

Abstract: This paper describes an original method for forming a stable coating on a polypropylene yarn. The use of this method provides this yarn with barrier antimicrobial properties, reducing its electrical resistance, increasing its strength, and achieving extremely high chemical resistance, similar to that of fluoropolymer yarns. The method is applied at the melt-spinning stage of polypropylene yarns. It is based on forming an ultrathin, continuous, and uniform coating on the surface of each of the yarn filaments. The coating is formed from polytetrafluoroethylene doped with magnetite nanoparticles stabilized with sodium stearate. The paper presents the results of a study of the effects of such an ultrathin polytetrafluoroethylene coating containing stabilized magnetite particles on the mechanical and electrophysical characteristics of the polypropylene yarn and its barrier antimicrobial properties. It also evaluates the chemical resistance of the polypropylene yarn with a coating based on polytetrafluoroethylene doped with magnetite nanoparticles.

Keywords: coatings; polypropylene yarn; polytetrafluoroethylene; magnetite nanoparticles; barrier antimicrobial properties; surface electrical resistance; chemical resistance; tensile strength

Citation: Prorokova, N.; Vavilova, S. Properties of Polypropylene Yarns with a Polytetrafluoroethylene Coating Containing Stabilized Magnetite Particles. *Coatings* **2021**, *11*, 830. https://doi.org/10.3390/coatings11070830

Academic Editor: Fabien Salaün

Received: 6 May 2021
Accepted: 6 July 2021
Published: 9 July 2021

Publisher's Note: MDPI stays neutral with regard to jurisdictional claims in published maps and institutional affiliations.

1. Introduction

Single-use materials (medical wear, masks, drapes and pads, sheets, etc.) are now often used in medical practice. A very important quality of such products is their antimicrobial properties, i.e., their ability to suppress the development of pathogenic microorganisms and protect the patients and doctors who come in contact with such microorganisms. One of the most widely applied methods of providing fibrous materials with antimicrobial properties is the use of metal and metal oxide nanoparticles [1–7] in antimicrobial preparations since they become attached to the surface of natural fibers with an enormous number of functional groups. Polypropylene (PP) fiber has a chemically inert smooth surface with no pores on it, and it is very hard to attach metal or metal oxide nanoparticles to the surface of such a fiber. However, it is known [8–14] that nanoparticles retain their antimicrobial properties even when they are immobilized inside this polymer. Therefore, PP fibers can also be provided with antimicrobial properties by immobilizing metal-containing nanoparticles in their inner cavities. A major problem with the introduction of nanoparticles into the polymer matrix during the melt spinning of fibers is that it is natural for nanoparticles to aggregate because even a slight aggregation of the fillers may adversely affect the strength of the fibers. It is rather difficult to prevent the aggregation of nanosized fillers during the melt spinning of nanomodified synthetic filaments because the formation of nanoparticle aggregates is the result of the metastability of the nanoparticles with excessive surface energy. The proposed solutions to this problem are mainly based on lowering the surface energy of nanosized fillers by treating their surfaces with special agents [14,15]. Different works [16–19] propose a different approach to solving this problem. This approach consists of the introduction of a polymer composite based on antimicrobial iron-, manganese-, and silver-containing nanoparticles stabilized by polyolefins into a polypropylene melt during yarn formation. When this method is applied, the nanoparticles are uniformly distributed throughout the fiber and are firmly kept within it. However, the stabilization method for metal-containing nanoparticles in their introduction into a polyolefin matrix during the synthesis procedure is rather complicated, which is an obstacle to their application in the preparation of modified PP yarns with antimicrobial properties. However, the use of unstabilized metal-containing nanoparticles for these purposes reduces yarn strength and, in case of large aggregate formation, leads to the clogging of the spinnerettes and yarn breakage.

The aim of this work was to attach nanoparticles firmly to the PP yarn surface instead of immobilizing them inside the yarn. In this way, we can exclude the negative effects of the aggregated nanoparticles on the strength of the yarn and can increase the antimicrobial effect by localizing the biologically active nanoparticles on the surface. To do this, we introduced nanoparticles into the thin polymer coatings formed on the yarns.

There are various methods of forming thin and elastic polymer coatings on fibrous materials for the direct change of their properties: in situ polymerization, vapor deposition, dipping in a solution, etc. [20–22]. We had earlier proposed a fundamentally new approach to obtaining PP yarns with a polytetrafluoroethylene (PTFE) coating [23–25]. According to this approach, PTFE adhesion to the surface of a thermoplastic yarn is achieved by depositing a suspension of finely dispersed PTFE on the surface of a semi-solidified yarn at the stage of its formation on the polymer (oiling stage). At the orientational drawing state, the coating then becomes much thinner due to the fluoroplastic pseudofluidity and high coefficient of thermal expansion, which causes the coating to become uniform and oriented. Such a continuous and uniform PTFE coating makes the yarn extremely chemoresistant, similar to a fluoroplastic yarn. We supposed that the introduction of a small number of antimicrobial metal-containing nanoparticles into the PTFE coating structure could provide the yarn with additional functional characteristics: the yarn could become antimicrobial, retaining its high chemoresistance. Moreover, the introduction of metal-containing nanoparticles into the coating structure might reduce the electrical resistance of the yarn surface. To achieve this effect, we introduced a controlled number of biologically active magnetite ($FeO \cdot Fe_2O_3$) nanoparticles into a PTFE suspension. To prevent the formation of large aggregates of magnetite nanoparticles, we stabilized them with a surfactant in advance. Since the orientational drawing of the yarn is carried out at temperatures up to 250 °C, we chose sodium stearate ($NaC_{18}H_{35}O_2$) for the magnetite stabilization, as preliminary studies had shown that it is a thermally stable surfactant. The obtained composition was deposited on a semi-solidified PP yarn and then subjected to orientational drawing.

2. Materials and Methods

The following materials and reagents were used in this work: granulated "Balen 01250" polypropylene with a melt index of 25 g/10 min and a melting point of 169 °C (Ufa, Ufaorgsintez, technical requirements No. 2211-015-00203521-99); sodium stearate ($NaC_{18}H_{35}O_2$), "pure"; ferrous sulfate heptahydrate ($FeSO_4 \cdot 7H_2O$), "chemically pure" (Moscow, Chimmed); iron trichloridehexahydrate ($FeCl_3 \cdot 6H_2O$), "pure" (Moscow, Chimmed); ammonium hydroxide (NH_4OH), 25%, "pure"; and a suspension of СФ-4Д(SF-4D) polytetrafluoroethylene (AO Galopolymer, Russia). In some of the experiments, we used a film made of 30 μm thick "Balen 01250" isotactic polypropylene (Europack-Ivanovo Ltd.» Ivanovo, Russia) as the PP yarn model.

The magnetite nanoparticles were prepared by codepositioning. For that purpose, we first prepared a solution of two salts containing 7.08 g $FeSO_4 \cdot 7H_2O$ (C = 0.5 M) and 3.75 g$FeCl_3 \cdot 6H_2O$ (C = 0.3 M), heated it to 80 °C and while mixing this solution, slowly added an excessive amount 15 mL (C = 1.5 M) of ammonia solution, NH_4OH, to it. The stabilization resulted in the production 1.0% $NaC_{18}H_{35}O_2$. Mixing the solution led to the appearance of black ultrafine particles. The mixture was then repeatedly washed with distilled water until the ammonia smell disappeared. The suspension of the stabilized magnetite was then dried in air until a powder was produced. The powder was sifted through a filter and dried for 24 h in a vacuum at a temperature of 60 °C.

The composition for coating the PP yarns was obtained from a finely dispersed suspension of SF-4D PTFE by mixing the components at a temperature of 80–90 °C. The composition contained a PTFE suspension-10%, sodium stearate-1.0%; magnetite-1%; and water-88%. To make the aggregates of the stabilized magnetite particles smaller, we subjected them to ultrasonic (US) treatment in a low-frequency ultrasonic sonicator of the USDN-2T type in a temperature-controlled container at a frequency of f = 22 kHz. The exposure lasted for 2 min.

The PP yarns were prepared using a laboratory bench for the SFPV-1 synthetic fibre spinning. The orientational drawing of the spun PP yarns was performed on an OSV-1 synthetic fiber orientation bench. Such benches can simulate the conditions of the industrial processes for yarn melt-spinning and orientational drawing. The images and schemes of the benches are presented in previous works [16,25].

The SFPV-1 spinning bench is equipped with an automated control panel for managing the spinning process, an extruder in which the polymer melts, a spinnerette with 24 holes ($\varnothing$ = 0.4 mm) for the formation of fine jets of liquid from the melt, godet wheels, and a fibre collecting drum for winding the spun yarn onto a spindle. During the experiment, the temperature values in the extruder zones were different: in the preheating zone, T_1 = 200 °C; in the melting zone, T_2 = 225 °C; in the melt stabilization, zone T_3 = 236 °C; and in the extrusion head heating zone, T_4 = 236 °C. The melt was fed at a rate of 20 g/min. The godet wheels operated at 100 m/min.

The composition containing a PTFE suspension and magnetite nanoparticles stabilized by a thermally stable surfactant was deposited on the PP yarn surface from the first and second godets at the oiling stage.

After the extrusion and deposition of the PTFE composition, the PP yarns were subjected to orientational drawing and were thermally stabilized on an OSV-1 bench. The process was carried out for the standard PP yarn at the following temperatures in the stretching zones: T_1 = 118–120 °C (the upper heated godet wheel), T_2 = 120–122 °C (the lower heated godet wheel), and T_3 = 123–125 °C (the thermoelectric plasticizer) at a rate of 3–20 m/min. The coated yarns were stretched at higher temperatures: T_1 = 120–135 °C, T_2 = 123–140 °C, and T_3 = 125–155 °C. Complex yarns composed of 24 filaments with a diameter of 15 µm were obtained. The coating thickness was 0.18 ± 0.5 µm. The characteristics of the PTFE coating are described in detail in [25,26].

In some of the experiments, the composition was deposited on the surface of a polypropylene film produced in industrial conditions from 30 µm thick "Balen" 01250 polypropylene (Europack, Ivanovo, Russia). The film became 5 times longer after being stretched on the OSV-1 bench at 120 °C.

The size of the magnetite particles in the suspensions and powders was determined with the Analysette 22 Compact, a laser light scattering particle size analyzer. The particle size range was 0.3–300 µm.

The basic mechanical characteristics of the PP yarns were determined by stretching the yarns once until breakage on a modernized 2099-P-5 tensile testing machine ("Tochpribor", OAO, Russia) in accordance with GOST 6611.2-73 (ISO 2062-72, ISO 6939-88).

The micrographs of the PP yarns were obtained using an optical microscope (equipped with a webcam (1.3 MP)) produced by "Biomed" (Russia). The surface structure was studied on a JSM 6380LA scanning electron microscope produced by JEOL.

The IR spectra were recorded on an Avatar ESP 360 type spectrometer (produced by the Nicolett company) with the method of multiple attenuated total reflectance (MATR) using a zinc selenide crystal with a 12-fold reflection in the range from 600 to 1600 cm^{-1}. This region contains the reflection band characteristics of PTFE and PP.

To make sure that the coating is resistant to abrasion, we used a PP film as a model of the PP yarn. The measurement was made on a PT-4 apparatus, a special machine for determining dye resistance to friction, according to the procedure described in [15]. The schematic diagram of the PT-4 apparatus is shown in Figure 1. The abrasive effect occurred during the simultaneous action of the normal applied load and the shear load produced by the application of a horizontal force. A film sample (1) was placed on the apparatus stage (2) and abraded with a calico piece fixed onto the protruding rubber stopper (3). The friction was the result of shifting the stage by 10 cm by pulling the stage handle (4) forward and backward a required number of times. The total force applied by the stopper to the stage was 9.8 N.

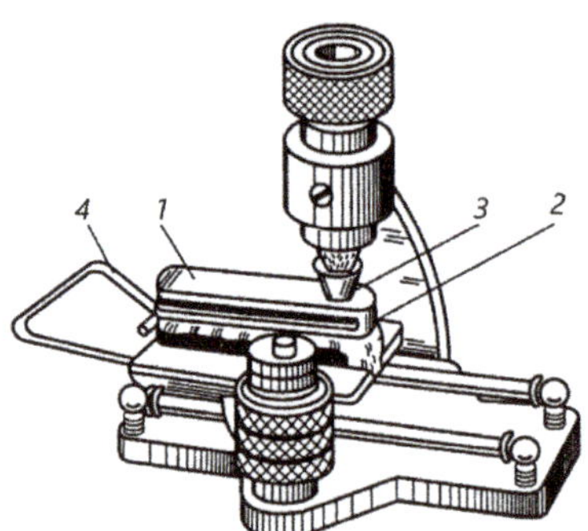

Figure 1. The schematic diagram of the PT-4 apparatus: *1*, a film sample; *2*, the apparatus stage; *3*, the rubber stopper; *4*, the stage handle.

The surface electrical resistance of the yarns (R) was determined with an IESN-1 apparatus, in which the measurements were made by a E6-13A teraohmmeter. The distinguishing feature of this apparatus is that one layer of the yarn is wound closely onto the sensor before the measurements. The sensor with a yarn wound onto it is mounted onto a dielectric support and is connected to the teraohmmeter. The electrical resistance measurements were made in accordance with ГОСТ 19806-74 (GOST 19806-74: a method of electric resistance determination using chemical threads). The specific surface electrical resistance was calculated by the formula:

$$\rho_s = \frac{kR}{\gamma^2 \sqrt{T\gamma}} \tag{1}$$

where k is a constant; for the IESN-1 apparatus, $k = 903.5 \ g^3/mm^8$;

T is the linear density of the yarns, tex;

γ is the yarn density, mg/mm^3;

R is the average electrical resistance value, ohm.

The effect of the modified fibrous material on the activity of pathogenic microorganisms was evaluated using typical test cultures: *Staphylococcus aureus* 6538-P ATCC=209-P FDA and *Escherichia coli*, strain M-17—Gram-positive and Gram-negative bacterial cultures, respectively, and *Candida albicans* CCM 8261 (ATCC 90028), a yeast-like microscopic fungi. A standard yarn sample was placed into a physiological solution containing a certain number of microbial colonies in the form of a suspension [27]. We kept the vials at room temperature for 24 h with constant shaking. The number of microbial colonies that the solution contained was determined by the changes in the solution transmission coefficient (the reference sample was assumed to have a 100% transmission coefficient), which was obtained by measuring the solution turbidity depending on the number of the colonies it contained. The reduction in the microbial contamination of the test objects relative to that of the reference object (saline) was evaluated in points: 1 point (0.0–0.1%) indicated that there was no antimicrobial effect; 2 points (0.1–90%)—a slight decrease in the number of microorganism colonies, an insufficient antimicrobial effect; 3 points (90–94%)—a significant reduction in the number of microorganism colonies, a good antimicrobial effect; 4 points (95–98%)—a significant reduction in the number of microorganism colonies, a very good antimicrobial effect; and 5 points (99% and higher)—a strong reduction in the number of microorganism colonies, an excellent antimicrobial effect.

The chemical resistance of a PP yarn with a PTFE coating was evaluated by measuring its tensile strength after it was exposed to aggressive liquids—a concentrated solution of sodium hydroxide (5 mol/L) at a temperature of 100 °C for 3 h and concentrated nitric acid (69%), also acting as a strong oxidizer, at a temperature of 25 °C for 24 h.

The resistance of a PP yarn with a PTFE coating to washing was assessed by the change in its specific tensile strength after repeated washings by being stirred in a solution of oleic soap (85%)—5 g/L and Na_2CO_3—2 g/L. The duration of each wash was 30 min, the temperature of each wash was 60 °C.

3. Results and Discussion

The supramolecular structure of melt-spun PP yarns is known to depend on the conditions of their spinning and orientational drawing [28–30]. Spinning leads to the formation of folded lamellar crystallites [30,31], which means that the resulting non-oriented yarns have extremely low strength and high (up to 1000%) elongation. At the orientational drawing stage, they become oriented, and the molecular chains in the lamellae completely unfold, forming fibrillar crystallites from the extended chains [32]. The better oriented the fibrils along the filament axes, the higher the yarn strength is. A yarn with a "perfect" and highly oriented structure and high strength can only be obtained by intensive orientational drawing. Its implementation, to a large extent, depends on the temperature and surface properties of the yarn. Depositing a PTFE coating on a PP yarn makes it possible to carry out its orientational drawing at higher temperatures than those used for a standard PP yarn. This makes the yarn strength parameters better [25,26]. However, it is necessary to find out how doping the PTFE coating with magnetite particles, namely the size of these particles, affects the yarn strength.

Smaller particles are known to have higher surface energy. This leads to the strong aggregation of the nanoparticles during their spinning and use. Such aggregates lead to the deterioration of a number of properties, in particular, they reduce the biological activity of nanosized particles. To avoid this, we stabilized the nanoparticles with a surfactant. When selecting the surfactant, we took into account that a PP yarn with a coating containing magnetite nanoparticles is subjected to orientational drawing at a temperature above 250 °C. For this reason, we chose sodium stearate, which is known to decompose at 300 °C and is resistant to oxidation. The stabilizer effect on the magnetite particle size and size distribution was determined based on the data given in Figure 2.

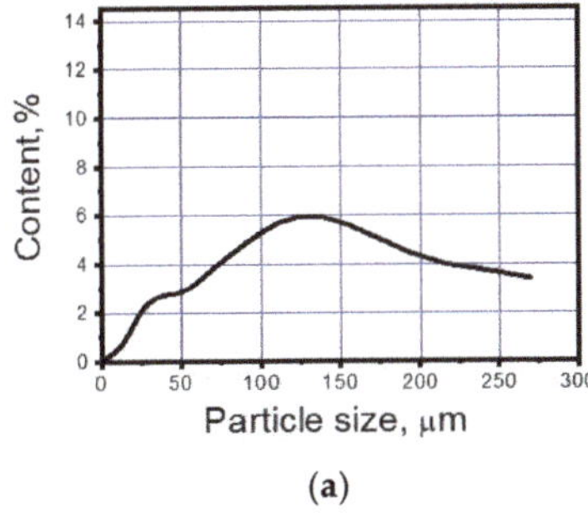
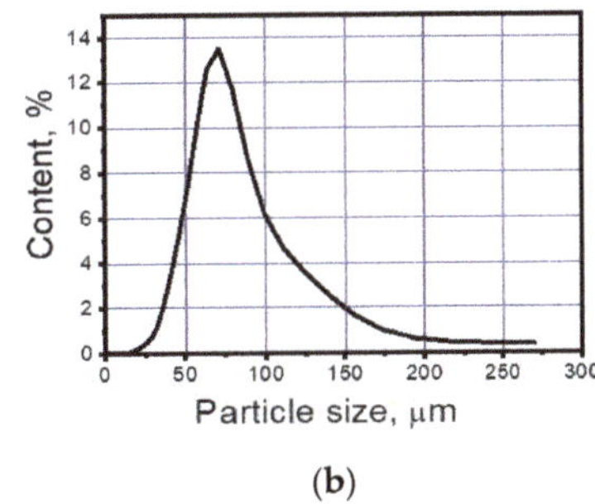

(**a**)   (**b**)

Figure 2. Size distribution of the magnetite particles synthesized (**a**) without a stabilizer; (**b**) using sodium stearate as the stabilizer.

It was found that the air drying of magnetite that is synthesized without a stabilizer leads to intensive particle aggregation. The measurement of the particle size distribution (Figure 2a) showed that the formed aggregates were up to 275 μm in size. Most of them were 150 μm in size. The use of sodium stearate as the stabilizer during the magnetite synthesis led to a considerable reduction (to 60–75 μm) in the average size of the aggregates that were formed (Figure 2b).

To make the aggregated particles of the magnetite even smaller, the magnetite suspension in water containing a surfactant was subjected to ultrasonic (US) treatment. The size distribution of the magnetite particles after the US treatment is shown in Figure 3.

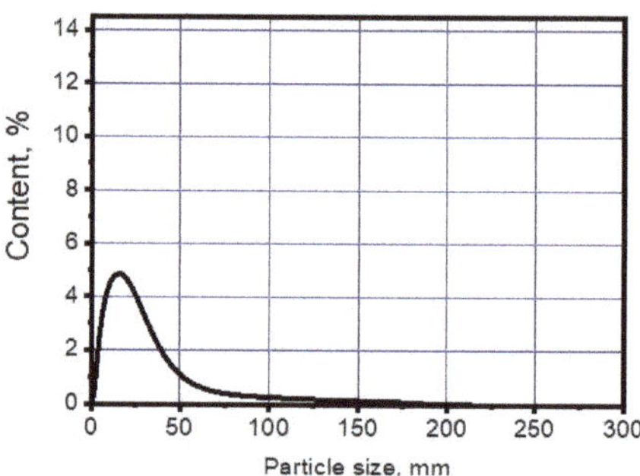

Figure 3. Size distribution after US treatment of magnetite particles synthesized using sodium stearate as the stabilizer.

Figure 3 shows that additional US treatment of the magnetite particles stabilized by sodium stearate reduces the average size of the aggregates to 15–25 μm, which can have a positive effect on the magnetite particle distribution in the PTFE coating.

To evaluate the effect of the magnetite particle sizes on the basic mechanical characteristics of the PP yarns, we measured the specific tensile strength and relative breaking elongation of the yarns coated with magnetite particles doped with non-stabilized and stabilized sodium stearate. For comparison, we also present the mechanical characteristics of the coated yarns after US treatment. The obtained data are shown in Table 1.

Table 1. Tensile strength and elongation of PP yarns with a PTFE coating doped with magnetite particles.

Components of the Composition, %			Specific Tensile Strength, MPa	Relative Breaking Elongation, %
PTFE	**Magnetite**	**Sodium Stearate**		
Standard PP yarn				
0	0	0	583 ± 23	33.6 ± 3.7
PP yarn with a PTFE coating				
10.0	0	0	643 ± 16	38.6 ± 2.7
PP yarn with a coating formed by a composition of PTFE and unstabilized magnetite				
10.0	1.0	0	466 ± 14	48.7 ± 4.4
PP yarn with a coating formed by a composition of PTFE and magnetite stabilized by sodium stearate				
10.0	1.0	1.0	578 ± 15	39.7 ± 4.5
PP yarn with a coating formed by a US-treated composition of PTFE and magnetite stabilized by sodium stearate				
10.0	1.0	1.0	658 ± 23	32.0 ± 1.7

Table 1 shows that doping coating with magnetite somewhat lowers the strength of the yarns. This phenomenon is associated with the fact that the composition contains magnetite aggregates that build into the coating structure, leading to the formation of microdefects in the coating. In the presence of a stabilizer, the strength reduction is not as great as it is without it, which is caused by smaller size of the magnetite aggregates. Ultrasonic treatment of the composition makes the magnetite aggregates even smaller. Consequently, the magnetite does not produce a negative effect on the strength of the yarn with a PTFE coating. A detailed analysis of the mechanical characteristics of the PP yarns with PTFE coating containing stabilized magnetite particles is presented in our previous work [33].

An additional confirmation of the fact that the composition after US treatment does not contain large magnetite aggregates is the micrographs shown in Figure 4. Figure 4 shows that the coating formed by the composition that has not been subjected to UV treatment (a) contains small aggregates (of up to several μm in size) of magnetite stabilized by sodium stearate. The small sizes of the aggregates are the result of the preliminary stabilization of the magnetite particles by sodium stearate. UV treatment of such composition (b) leads to aggregate destruction. The resulting coating has a uniform structure without any inclusions.

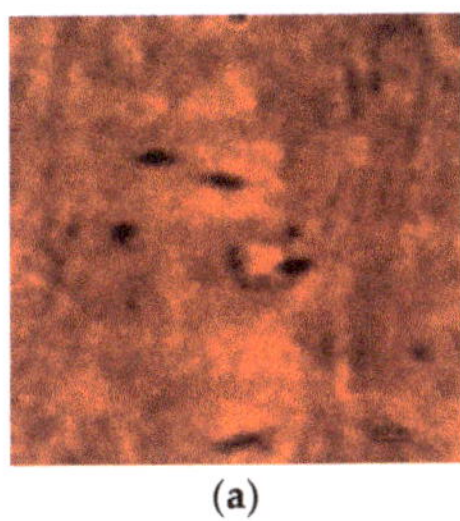 

(a) (b)

Figure 4. Micrographs of the film with a coating formed by a composition of PTFE and stabilized magnetite: (**a**) without US treatment; (**b**) after US treatment. Optical microscopy method. The magnification is 1000 times.

Figure 5 shows a scanning electron microscopy image of the coating containing PTFE and magnetite stabilized by sodium stearate formed by the UV-treated composition. The image also indicates that the coating structure is uniform and does not have any noticeable foreign inclusions.

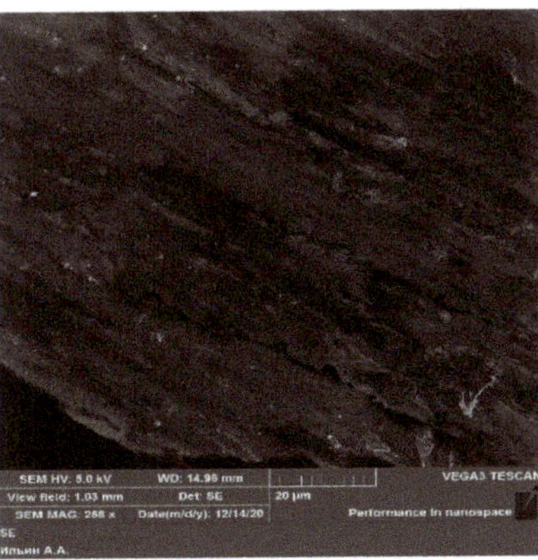

Figure 5. Image of the film with a coating formed by a US-treated composition of PTFE and magnetite stabilized by sodium stearate.

Materials made from PP and PTFE are known to become strongly electrified while being processed and utilized. That is why we assumed that the introduction of even a small number of conductive magnetite particles into a PTFE coating could reduce the surface electrical resistance of PP yarns and films and, consequently, lower their ability to become electrified.

Table 2 shows the results of the surface electrical resistance measurement in the PP films with the PTFE coatings containing magnetite.

Table 2. Surface electrical resistance of PP yarns with a PTFE coating doped with magnetite particles.

Components of the Composition, %			Surface Electrical Resistance, Ohm
PTFE	**Magnetite**	**Sodium Stearate**	
Standard PP yarn			
0	0	0	$4.5 \cdot 10^{14}$
PP yarn with a PTFE coating			
10.0	0	0	exceeds the apparatus measurement range
PP yarn with a coating formed by a composition of PTFE and unstabilized magnetite			
10.0	1.0	0	$3.4 \cdot 10^{14}$
PP yarn with a coating formed by a composition of PTFE and magnetite stabilized by sodium stearate			
10.0	1.0	1.0	$3.8 \cdot 10^{10}$
PP yarn with a coating formed by a US-treated composition of PTFE and magnetite stabilized by sodium stearate			
10.0	1.0	1.0	$5.7 \cdot 10^{8}$

As Table 2 shows, the inclusion of large unstabilized magnetite particles in the PTFE coating structure does not lower the electrical resistance of the film structure. At the same time, the films with a PTFE coating doped with stabilized magnetite particles have a much lower surface electrical resistance than the PP films and films with an undoped PTFE coating.

An important thing to note is that the minimum surface electrical resistance is characteristic of films with a coating formed by a US-treated composition, i.e., the one containing magnetite particles of the minimum size. This composition was used in the further experiments.

One of the most important properties of the PP yarns with a PTFE coating doped with magnetite nanoparticles is their ability to suppress the activity of pathogenic microorganisms. Stabilized magnetite particles build themselves into the coating structure and cannot diffuse outside of it. That is why we evaluated the antimicrobial properties of the coated yarns using the calculation procedure normally used for determining the antimicrobial properties of nonmigrating preparations [27]. The reduction in microbial contamination of the test objects in comparison to the same indicator in the reference object (physiological solution) was evaluated in points. The results of the evaluation of the activity of the PP yarn with a PTFE coating containing magnetite particles against the test Gram-positive and Gram-negative bacteria are shown in Table 3.

Table 3 shows that the formation of a PTFE coating with high anti-adhesion characteristics provides the PP yarn with weak antimicrobial properties. The introduction of magnetite nanoparticles into the coating structure significantly strengthens the antimicrobial activity of the yarn. The coated yarns considerably reduce the number of pathogenic bacterial communities, i.e., the yarns with a PTFE coating containing magnetite particles exhibit excellent antibacterial activity against *Escherichia coli*—a type of Gram-negative bacteria—and *Staphylococcus aureus*—a type of Gram-positive bacteria. The yarn also produces a satisfactory inhibiting effect on the activity of the *Candida albicans* microfungi. It should be said that the antimicrobial action of a modified yarn is triggered by its direct contact with microorganisms, i.e., PP fibrous materials with a PTFE coating containing stabilized magnetite particles possess barrier antimicrobial properties.

Table 3. Antimicrobial properties of a PP yarn with a PTFE coating containing stabilized magnetite particles.

PP Yarn Type	Inhibition of Activity of Pathogenic Microorganisms, Points/%		
	Escherichia coli	*Staphylococcus aureus*	*Candida albicans*
Standard PP yarn	2/39	2/46	2/41
PP yarn with a PTFE coating	3/90	3/93	3/90
PP yarn with a PTFE coating containing 1.0% of stabilized magnetite	5/99	4/97	3/94

As is known, during its formation, processing, and role in goods production, a yarn is subjected to severe mechanical effects. One of the most important effects is aging. That is why it is necessary to evaluate the coating resistance to abrasion.

Since a single filament of a PP yarn has a small diameter (~15 μm) and its surface is characterized by a great curvature, the PP substrate interaction with a PTFE coating was studied on a model object—a PP film with a PTFE coating doped with stabilized magnetite. To study the coating resistance to abrasion, we subjected the films to abrasion on a PT-4 apparatus. The residual amounts of the coating on the film were determined by the presence of intensive bands associated with the CF_2 group vibrations (1211 and 1154 cm^{-1}) in the spectra [34], using the IR-spectroscopy method (MATR). Figure 6 shows the IR-spectra (MATR) of the PP films with PTFE coatings doped with magnetite stabilized by sodium stearate after subjecting the films to abrasion a different number of times.

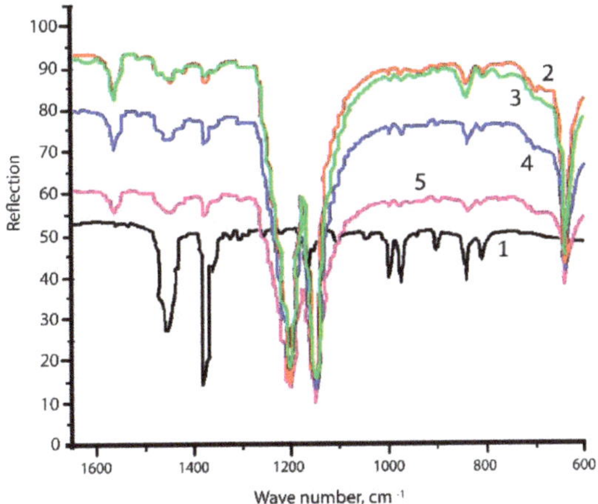

Figure 6. IR-spectra (MATR) of PP films with PTFE coatings doped with magnetite stabilized by sodium stearate and subjected to abrasion a different number of times: 1, without a coating; 2, with a coating; 3, with a coating subjected to abrasion 10 times; 4, with a coating subjected to abrasion 50 times; 5, with a coating subjected to abrasion 500 times.

As Figure 6 shows, the spectra of all of the films subjected to abrasion have bands that are characteristic of PTFE. Figure 5 shows that the magnetite particles are built into the coating structure and make up its integral parts. Thus, it can be concluded that even after intensive abrasion, the films retain their fluoroplastic coating doped with stabilized magnetite. Applying the baseline method, we calculated the PTFE content on the film surface subjected to abrasions of different intensities. The internal standard for isotactic polypropylene was the band at 1460 cm^{-1} [35], and for measuring the PTFE content—the band at 1211 cm^{-1}. The results of the calculation are given in Table 4.

Table 4. Effects of abrasion on the PTFE content on the surface of an oriented PP film (according to the IR spectra).

Number of Abrasive Effects, Cycle.	Ratio of the Height of the Band at 1211 to That at 1450 cm^{-1}
Film without a coating	0.1
0	11.9
10	11.4
50	11.0
250	8.5
500	8.4

An analysis of the data presented in Table 4 shows that the coating is highly resistant to abrasion. Even after 500 cycles of abrasive action, the fluoroplastic layer thickness becomes only slightly thinner.

To determine whether the PP yarn with a PTFE coating retained its extremely high chemical resistance after the introduction of a small amount of stabilized magnetite into the coating structure, we measured the changes in its tensile strength after long-time exposure of the yarns to aggressive liquids—concentrated solutions of sodium hydroxide and nitric acid, with the latter acting as a strong oxidizer at the same time. The obtained data are given in Table 5.

Table 5. Tensile strength and elongation of PP yarns with a PTFE coating after boiling in a sodium hydroxide solution (5 mol/L) for 3 h and storage in concentrated nitric acid for 24 h at a temperature of 25 °C.

Without Treatment		After Boiling in a NaOH Solution (Concentrated)		After Storage in HNO$_3$ (Concentrated)	
Specific Tensile Strength, MPa	Relative Breaking Elongation, %	Specific Tensile Strength, MPa	Relative Breaking Elongation, %	Specific Tensile Strength, MPa	Relative Breaking Elongation, %
Standard PP yarn					
583 ± 23	33.6 ± 3.7	539 ± 21	52.0 ± 4.8	428 ± 22	40.4 ± 4.1
PP yarn with a PTFE coating					
643 ± 16	38.6 ± 2.7	728 ± 19	44.9 ± 3.3	748 ± 19	27.0 ± 2.9
PP yarn with a coating formed by a US-treated composition of PTFE and magnetite stabilized by sodium stearate					
658 ± 23	32.0 ± 1.7	720 ± 17	45.2 ± 3.1	761 ± 25	24.3 ± 1.7

Table 5 shows that the strength of a standard PP yarn subjected to the action of concentrated alkalis and acids becomes 8–27% lower. The effect of aggressive liquids on the PP yarn with a PTFE coating does not lead to a loss in strength. Moreover, it makes the strength higher. As work [25] shows, the higher strength of the yarn after its exposure to chemically aggressive liquids is the result of removing the excessive amount of PTFE that is weakly bound to the PP substrate from it. As a result, it makes the coating structure even more uniform and its surface smoother. The chemical resistance of PP yarns with a PTFE coating doped with stabilized magnetite is not lower than that of PP yarns with an undoped PTFE coating. Thus, a PP yarn with a PTFE coating doped with magnetite retains its extremely high resistance.

PP yarn with a PTFE coating, and therefore products made from it, have a very high resistance to washing. Tests have shown that the specific tensile strength of such yarns remains unchanged after 40 washes that are 30 min in duration.

4. Conclusions

The paper proposes an original method for forming a coating on a PP yarn surface, which provides the yarn with a number of new working characteristics. The novelty of this approach is the formation a stable ultrathin coating of PTFE doped with magnetite nanoparticles stabilized on the surface of a PP yarn by a thermally stable surfactant. The PTFE in the coating makes it possible to carry out orientational drawing at temperatures exceeding the standard ones, which considerably increases the strength of the yarn. The doping of the coating with magnetite nanoparticles possessing strong antimicrobial activity and electric conductivity provides the yarn with barrier antimicrobial properties and reduces the surface electrical resistance. A yarn with a PTFE coating doped with stabilized magnetite particles also exhibits extremely high chemical resistance similar to that of fluoropolymer yarns. The coating is characterized by a high resistance to abrasion, which means that it is durable.

All of the enumerated properties of the yarn with a PTFE coating doped with stabilized magnetite nanoparticles make it suitable for preparing interior materials that can be used in public transport and in other places with large gatherings of people, including those with weak immunity, physical development problems, elderly people as well as in kindergartens, nursing homes, and hospitals.

Author Contributions: Conceptualization, methodology, writing—original draft preparation, writing—review and editing, N.P.; data curation, software, investigation, visualization, S.V. All authors have read and agreed to the published version of the manuscript.

Funding: This work was financially supported by the Ministry of Science and Higher Education of the Russian Federation (state contract No. 01201260484).

Institutional Review Board Statement: Not applicable.

Informed Consent Statement: Not applicable.

Acknowledgments: The authors would like to thank O.Yu. Kuznetsov (Ivanovo State Medical Academy) for his assistance in the microbiological studies and I.V. Kholodkov (Ivanovo State University of Chemistry and Technology) for his assistance in the scanning electron microscopy studies. The work was conducted using equipment belonging to the Center for Joint Use of Scientific Equipment at "The Upper Volga Region Center of Physico-Chemical Research" and the Center for Joint Use of Ivanovo State University of Chemistry and Technology.

Conflicts of Interest: The authors declare no conflict of interest.

References

1. Nadtochenko, V.F.; Radtsig, M.A.; Khmel, I.A. Antimicrobial effect of metallic and semiconductor nanoparticles. *Nanotechnol. Russ.* **2010**, *5*, 277–289. [CrossRef]
2. Palza, H. Antimicrobial Polymers with Metal Nanoparticles. *Int. J. Mol. Sci.* **2015**, *16*, 2099–2116. [CrossRef] [PubMed]
3. Kon, K.; Rai, M. Metallic nanoparticles: Mechanism of antibacterial action and influencing factors. *J. Comp. Clin. Path. Res.* **2013**, *2*, 160–174.
4. Wang, H.; Zakirov, A.; Yuldashev, S.U.; Lee, J.; Fu, D.; Kang, T. ZnO films grown on cotton fibers surface at low temperature by simple two-step process. *Mater. Lett.* **2011**, *65*, 1316–1327. [CrossRef]
5. Borkow, G.; Gabbay, J. Copper, an ancient remedy returning to fight microbial and viral infections. *Curr. Chem. Biol.* **2009**, *3*, 272–278. [CrossRef]

6. Medici, S.; Peana, M.; Nurchi, V.M.; Zoroddu, M.A. Medical uses of silver: History, myths, and scientific evidence. *J. Med. Chem.* **2019**, *62*, 5923–5943. [CrossRef]

7. Afraz, N.; Uddin, F.; Syed, U.; Mahmood, A. Antimicrobial finishes for textiles. *Curr. Trends Fashion Technol. Text. Eng.* **2019**, *4*, 555646. [CrossRef]

8. Delgado, K.; Quijada, R.; Palma, R.; Palza, H. Polypropylene with embedded copper metal or copper oxide nanoparticles as a novel plastic antimicrobial agent. *Lett. Appl. Microbiol.* **2011**, *53*, 50a–54a. [CrossRef]

9. Palza, H.; Gutie'rrez, S.; Delgado, K.; Salazar, O.; Fuenzalida, V.; Avila, J.I.; Figueroa, G.; Quijada, R. Toward Tailor-Made Biocide Materials Based on Poly(propylene)/Copper Nanoparticles. *Macromol. Rapid Commun.* **2010**, *31*, 563–567. [CrossRef]

10. Ziąbka, M.; Dziadek, M. Long lasting examinations of surface and structural properties of medical polypropylene modified with silver nanoparticles. *Polymers* **2018**, *11*, 2018. [CrossRef]

11. Mania, S.; Cie'slik, M.; Konzorski, M.; Swiecikowski, P.; Nelson, A.; Banach, A.; Tylingo, R. The synergistic microbiological effects of industrial produced packaging polyethylene films incorporated with zinc nanoparticles. *Polymers* **2020**, *12*, 1198. [CrossRef]

12. Li, Q.-S.; He, H.-W.; Fan, Z.-Z.; Zhao, R.-H.; Chen, F.-X.; Zhou, R.; Ning, X. Preparation and performance of ultra-fine polypropylene antibacterial fibers via melt electrospinning. *Polymers* **2020**, *12*, 606. [CrossRef] [PubMed]

13. Gawish, S.M.; Mosleh, S. Antimicrobial polypropylene loaded by silver nano particles. *Fibers Polym.* **2020**, *21*, 19–23. [CrossRef]

14. Zhang, G.; Xiao, Y.; Yan, J.; Xie, N.; Liu, R.; Zhang, Y. Ultraviolet light-degradation behavior and antibacterial activity of polypropylene/ZnO nanoparticles fibers. *Polymers* **2019**, *11*, 1841. [CrossRef] [PubMed]

15. Geller, V.E. Prospects for preparing nanocomposite textile yarn (review). *Fibre Chem.* **2014**, *45*, 65–70. [CrossRef]

16. Prorokova, N.P.; Vavilova, S.Y.; Biryukova, M.I.; Yurkov, G.Y.; Buznik, V.M. Modification of polypropylene filaments with metal containing nanoparticles immobilized in a polyethylene matrix. *Nanotechnol. Russ.* **2014**, *9*, 533–540. [CrossRef]

17. Prorokova, N.P.; Vavilova, S.Y.; Kuznetsov, O.Y.; Buznik, V.M. Antimicrobial properties of polypropylene yarn modified by metal nanoparticles stabilized by polyethylene. *Nanotechnol. Russ.* **2015**, *10*, 732–740. [CrossRef]

18. Prorokova, N.P.; Vavilova, S.Y.; Biryukova, M.I.; Yurkov, G.Y.; Buznik, V.M. Polypropylene threads modified by iron-containing nanoparticles stabilized in polyethylene. *Fibre Chem.* **2016**, *47*, 384–388. [CrossRef]

19. Prorokova, N.P.; Buznik, V.M. New methods of modification of synthetic fibrous materials. *Rus. J. Gen. Chem.* **2017**, *87*, 1371–1377. [CrossRef]

20. Urvashi, M.; Subhankar, M.; Arobindo, C. Polypyrrole-silk electro-conductive composite fabric by in situ chemical polymerization. *J. Appl. Polym. Sci.* **2015**, *132*, 41336–41345. [CrossRef]

21. Maity, S.; Chatterjee, A.; Singh, B.; Pal Singh, A. Polypyrrole based electro-conductive textiles for heat generation. *J. Text. Inst.* **2014**, *105*, 887–893. [CrossRef]

22. Villanueva, R.; Ganta, D.; Guzman, C. Mechanical, *in-situ* electrical and thermal properties of wearable conductive textile yarn coated with polypyrrole/carbon black composite. *Mater. Res. Express.* **2019**, *6*, 016307. [CrossRef]

23. Prorokova, N.P.; Vavilova, S.Y.; Kumeeva, T.Y.; Moryganov, A.P.; Buznik, V.M. Synthetic Yarns with High Chemical Resistance and Low Coefficient of Friction. RU Patents 2522337, 10 July 2014.

24. Prorokova, N.P.; Vavilova, S.Y.; Kumeeva, T.Y.; Moryganov, A.P.; Buznik, V.M. Method for Producing Synthetic Yarns. RU Patents 2522338, 10 July 2014.

25. Prorokova, N.P.; Vavilova, S.Y.; Bouznik, V.M. A novel technique for coating polypropylene yarns with polytetrafluoroethylene. *J. Fluorine Chem.* **2017**, *204*, 50–58. [CrossRef]

26. Prorokova, N.P.; Vavilova, S.Y. Bulk and surface modification of polypropylene filaments at the stage of their formation from a melt. *Fibre Chem.* **2018**, *50*, 233–238. [CrossRef]

27. ASTM E2149—10 Standard Test Method for Determining the Antimicrobial Activity of Immobilized Antimicrobial Agents Under Dynamic Contact Conditions. USA. 2001. Available online: https://standards.globalspec.com/std/14338807/ASTM%20E2149 (accessed on 8 July 2021).

28. Fujiyama, M.; Wakino, T.; Kawasaki, Y.J. Structure of the skin layer in injection molded polypropylene. *Appl. Polym. Sci.* **1988**, *35*, 29–49. [CrossRef]

29. Kolb, R.; Seifert, S.; Stribeck, N.; Zachmann, H.G. Simultaneous measurements of small- and wide-angle X-ray scattering during low speed spinning of poly(propylene) using synchrotron radiation. *Polymer* **2000**, *41*, 1497–1505. [CrossRef]

30. Zavadskii, A.E.; Vavilova, S.Y.; Prorokova, N.P. X-ray Analysis of the texture of freshly spun polypropylene threades. *Fiber Chem.* **2013**, *45*, 145–149. [CrossRef]

31. Zavadskii, A.E.; Vavilova, S.Y.; Prorokova, N.P. Orientation processes in crystalline and amorphous regions of polypropylene during yarn spinning. *Fibre Chem.* **2017**, *49*, 10–14. [CrossRef]

32. Zavadskii, A.E.; Vavilova, S.Y.; Prorokova, N.P. X-Ray diffraction of the supramolecular structure of polypropylene yarns. *Fiber Chem.* **2014**, *46*, 222–227. [CrossRef]

33. Prorokova, N.P.; Vavilova, S.Y.; Bouznik, V.M. Mechanical characteristics of polytetrafluoroethylene coated polypropylene yarns made by new technology. *Khimicheskaya Tekhnologiya* **2020**, *9*, 409–417. [CrossRef]

34. Ignateva, L.N.; Buznik, V.M. IR-spectroscopic examination of polytetrafluoroethylene and its forms. *Rus. J. Gen. Chem.* **2009**, *79*, 677–685. [CrossRef]

35. Dechant, J.; Danz, R.; Kimmer, W.; Schmolke, R. *Ultrarotspektroskopische Untersuchungen an Polymeren*; Akademie-Verlag: Berlin, Germany, 1972.

MDPI
St. Alban-Anlage 66
4052 Basel
Switzerland
Tel. +41 61 683 77 34
Fax +41 61 302 89 18
www.mdpi.com

Coatings Editorial Office
E-mail: coatings@mdpi.com
www.mdpi.com/journal/coatings

www.ingramcontent.com/pod-product-compliance
Lightning Source LLC
LaVergne TN
LVHW070614170726
843515LV00004B/72